John Ellor Taylor

The Aquarium

Its Inhabitants, Structure & Management

John Ellor Taylor

The Aquarium
Its Inhabitants, Structure & Management

ISBN/EAN: 9783337267056

Printed in Europe, USA, Canada, Australia, Japan

Cover: Foto ©berggeist007 / pixelio.de

More available books at **www.hansebooks.com**

THE AQUARIUM;

ITS INHABITANTS, STRUCTURE, AND MANAGEMENT.

THE AQUARIUM;

ITS INHABITANTS, STRUCTURE, AND MANAGEMENT.

BY

J. E. TAYLOR, Ph.D., F.L.S., F.G.S., Etc.

AUTHOR OF 'HALF-HOURS IN THE GREEN LANES,'
'HALF-HOURS AT THE SEA-SIDE,' 'GEOLOGICAL STORIES,' ETC.

LONDON:

HARDWICKE & BOGUE, 192, PICCADILLY, W.

1876.

LONDON: PRINTED BY WILLIAM CLOWES AND SONS, STAMFORD STREET AND CHARING CROSS.

PREFACE.

We regard the institution of public aquaria as more or less the result of the deeper interest now felt in the life-histories of aquatic animals, consequent upon that extensive knowledge of natural history which is one of the intellectual features of our time. We believe their extension will be greater, on this account, than those people imagine who hold they will share the fate of "spelling bees," &c. That they are a popular means of education none will deny, and the success they have everywhere met with leads us to hope they are serving a good purpose.

This little volume is intended as a handbook or popular manual to our public aquaria, so as to render them still more effective as a means of education. Their history, construction, and principles of management have been briefly described, as also the natural history of the chief animals which have been more or less successfully acclimatised. The list of the latter is constantly being extended, and there appears no

limit to the number which may be healthily maintained and exhibited.

We have had the great advantage of having the
following pages overlooked by Mr. W. A. Lloyd, of
the Crystal Palace, to whom our best thanks are due
for many valuable suggestions.

The work is now presented to the public in the
hope that it may add to the educational effects of
our public aquaria, and be the means of rendering
the education in zoology more popular and extensive.

IPSWICH, *September* 21, 1876.

CONTENTS.

CHAPTER I.

THE HISTORY OF AQUARIA.

CHAPTER II.

HISTORY OF AQUARIA—*continued.*

CHAPTER III.

PRINCIPLES OF THE AQUARIUM.

CHAPTER IV.

CONSTRUCTION OF FRESH-WATER AQUARIA.

CHAPTER VII.

MOLLLUSCA, INSECTS, ETC., OF THE FRESH-WATER AQUARIUM.

CHAPTER VIII.

THE AQUARIUM AS A NURSERY FOR THE MICROSCOPE.

CONTENTS.

CHAPTER X.

OUR PUBLIC AQUARIA.

CHAPTER XI.

MAMMALIA, REPTILIA, AND FISHES OF PUBLIC MARINE AQUARIA.

CHAPTER XIV.

CUTTLE-FISH, MOLLUSCA, ETC., OF MARINE AQUARIA.

CHAPTER XV.

CRUSTACEA, ECHINODERMS, ANNELIDS, ETC., OF MARINE AQUARIA.

CHAPTER XVI.

SEA-ANEMONES AND OTHER ZOOPHYTES, ETC., OF MARINE AQUARIA.

THE AQUARIUM.

CHAPTER I.

THE HISTORY OF AQUARIA.

NOTWITHSTANDING a good deal of quibbling which
has taken place respecting the word "Aquarium,"
there can be no doubt it has now gained its ground,
as signifying contrivances for the support of living
fresh-water and marine animals under such artificial
conditions as resemble their natural surroundings.
The word has passed out of the region of philology
into that of common parlance, and has now become
stereotyped in dictionaries.

The charming works of P. H. Gosse undoubtedly
did much to make aquarium keeping popular about
twenty-three years ago. Everyone who loved nature
could not help feeling attracted towards the lovely
objects, which he showed were so abundant on our
coast, after his animated descriptions of them. To
a great extent this was in advance of the natural
science of the time, and although it was the means

of collecting a great deal of information relative to the habits of the invertebrate animals, it had to fall back until science came up with it. The enormous strides which natural science has made since the publication of the 'Origin of Species' have necessitated large aquaria, where the new study of the embryology and larval conditions of the lower animals could be more easily followed. Since that time, also, zoology has become more attractive even to general readers. The fact that evolutionists and non-evolutionists have taken sides over zoological questions, renders it imperative that both shall observe more and theorise less. It has been found, also, that large aquaria may be rendered places of the highest amusement, as well as of the easiest and pleasantest instruction. Hence their numbers are largely increasing, and we doubt not the time is not far distant when all our large towns will be provided with them, so that all classes may know more of the marvellous works of God. To economists, aquaria cannot fail to be of the highest interest, for even within the last few years, observation at several of them has settled various most important facts relating to the life-history of some of those creatures which are most valuable to us as food. In one instance, at least, it was the means of preventing the framing of a law that was based on zoological ignorance, and which would have done as much harm to our fish supply as it was intended to do good ! In 1865 a Royal Commission, on which several naturalists sat, met at some of our fishing ports, and took evidence from fishermen and others as to whether trawling did not do much harm, by breaking up the sea bed where the ova of fish had been deposited. The idea then was that the cod and whiting—two of the most abundant of our native food-fishes — deposited their eggs on the sea floor. Professor Sars, the well-known Danish naturalist, had expressed his opinion that the ova of these fish floated on the surface; but it was first substantiated in the Brighton Aquarium, where it was found that the ova both of these fish and the mackerel, floated on the surface during the entire period of their development. Had it not been for this discovery, it is more than likely that by this time the fishing trade, as well as the fish supply, would have been crippled by a law which would have restrained trawling operations over cod grounds during the whole of the spawning season.

Mr. Saville-Kent, at the Manchester Aquarium, has contributed towards the history of the common herring, from its young state. Dr. Gunther, the well-known ichthyologist, had already declared his belief that by far the greater part of " whitebait" consisted of the fry of herrings.* If this is so, then, in consuming them so recklessly, we are interfering with the chief fish-food of the common people. Mr. Lloyd

* Dr. Gunther affirmed that *Clupea alba*, or whitebait, was the young of *Clupea harengus*, or the common herring.

(who was the first to keep whitebait in London, in 1858), and others had experienced a difficulty in herring culture. In consequence of the migratory habits of these fish they often injured themselves by dashing against the glass or the rockwork of the tank in which they were kept. As they move about principally by night, it struck Mr. Kent to illuminate the tank by a feeble light, so that the outlines of the walls, rocks, &c., should be visible to the fish. This plan succeeded admirably, and by its aid Mr. Kent kept and fed whitebait until they have grown to half the size of the ordinary herring. At the time he made this announcement, the fish were eighteen months old.*

In addition to the above-mentioned important facts with which our large aquaria have made us acquainted, there are others not yet worked out, but which are in process of careful observation. It was discovered in the Hamburg Aquarium that the *Phyllosoma*, one of the "glass crabs," which had been placed in a separate order prepared for it, is only the young of the crawfish (*Palinurus quadricornis*). The Brighton Aquarium has further contributed important information as to the rapidity of the growth of the salmon. Before then, the growth of this fish was thought to be much slower than observation and ex-

---

* Mr. Kent obtained his specimens in a very young state. The only person who has hatched out herrings in aquaria is Mr. Stephenson, of Brixton.

periment have proved. At the Southport Aquarium, under Mr. C. L. Jackson, experiments are being conducted which will make us better acquainted with the life-history of another valuable food-fish, the salmontrout.

Although artificial contrivances for preserving fish alive have undoubtedly been in vogue for many centuries, aquaria, in the sense in which we understand the word, are peculiarly modern. The ancient Romans paid as great attention to their fish-ponds as wealthy gentlemen, of horticultural tastes, now do to their orchid and fern houses. No expense seems to have been spared in making these fish-ponds as large and attractive as possible, or in obtaining valuable and beautiful fish for stocking them. Amongst others, the red mullet (*Mullus barbatus?*) appears to have been the greatest favourite. It was kept in the ponds for the sake of its beauty, and was usually brought to the table alive, so that the assembled guests could indulge in the pleasure of witnessing the rapidly changing prismatic tints which the fish assumed whilst dying. Not unfrequently canals led from the fish-ponds into the banqueting hall. The red mullet, when it attained a large size, was of great value; one of four pounds and a half fetching a sum equal to 60*l.* sterling. These mullets are immortalised by the price that was given for them in the reign of Caligula, about 240*l.* Pliny relates that the fish-pond of one of the Roman patricians (C. Herius) was sold for a sum amounting to more than 32,000*l.* So extensive were these ponds, and so well stocked, that the same gossipy naturalist tells us the fish alone from the ponds of Lucullus, the well-known gourmand, fetched a sum as large as that just named! The Romans were capital judges of another modern delicacy, the oyster, the modern demand for which has been run almost as high as it was nearly two thousand years ago. Reservoirs were constructed for the preservation of oysters, and large sums of money were laid out in getting stock and taking proper care of them.

The Chinese have long kept live fish for the table and market. Our well-known gold and silver fish (*Cyprinus auratus*) come from their country, and were introduced into Europe as ornamental living objects more than two centuries ago. Pepys perhaps refers to these in his ' Diary,' as a " fine rarity ; of fishes kept in a glass of water, that will live so for ever— and finely marked they are, being foreign." Both the Japanese and Chinese have long kept these fish in artificial tanks and glasses for amusement, and have succeeded in roughly training them. During the middle ages, fish-ponds were esteemed a necessary appurtenance to monasteries, abbeys, and even halls. The long abstinence from all animal food, except fish, during Lent, and the many other fasting days imposed by the Church, rendered it necessary that fish of some sort should be easily available for use. The moats which ran round castles or other baronial buildings, often served the double purpose of defence and fish preserves. In the immediate neighbourhood of abbeys we usually find large fish-ponds, unless (as is frequently the case) these religious buildings stood near some well-known stream. No doubt at this time, in spite of the difficulty of transit, European fish were more or less interchanged, so that it does not do to accept their present geographical distribution as a natural one. It is all but certain that the carp was brought from southern Europe to the more northerly parts ; its great size and esteemed flavour rendering it a favourite. The pike is said to have been introduced into England in like manner, but this is hardly likely, as we find its remains in the post-glacial river-bed of Mundesley, in Norfolk. Thus we have incontestable evidence of the existence in Britain of the pike long before the historic period, and when the physical geography of the surface was, in Norfolk at least, very different from what it is now. Tastes, as regards fish and other aquatic animals, have differed much since mediæval times. The upper classes regarded pike and tench as fit only for the lower orders, whilst they did not scruple to enjoy the coarse flesh of the sea dog, the porpoise, and even the whale! In an old document of the thirteenth century, about fifty kinds of fish are mentioned which were retailed in the French markets. Lacroix says that a century later, the flesh of the whale was salted down for the use of the common people. Congers,

cuttle-fish, and sturgeon were the principal food-fishes
of the masses ; whilst turbot, sole, and " John Dory"
had even then obtained, by their high price, the
aristocratic position of catering only for the stomachs
of the wealthy.

The edible frog (*Rana esculenta*) is another animal
which has been specially cared for by those who
have learned to like it as an article of food. Tanks
or ponds, in which it can pass through its ordinary life-history, and whence it can easily be fished
out for the table, still exist in France. Of course
we need not here do more than remark that the
edible frog is another species than that which is so
common in England ; although there is no reason
in the world why the latter should not be as dainty
an article of food, if there were only more of it. Pond
frogs were regarded as among their choicest morsels
by the ancient Gauls and Franks, in whose country
these amphibians have continued to be more or less
favourites ever since. Formerly they were served at
the best tables, dressed with a green sauce.

Between the artificial contrivances for the preservation of aquatic and other animals designed for the
table, and the modern aquaria in which they are kept
to administer to the growing love for knowledge,
there is as great a gulf fixed as there is between the
mind and the stomach. Very little knowledge indeed
has been handed down to us from the costly piscinæ
of the ancient Romans, or the more homely fish-

ponds of mediæval times. From the lofty eminence whence ignorant men looked down, all lowlier creatures seemed beneath their study. It remained for the era when we had learned to regard all things that God has made as worthy of our consideration, to increase our knowledge of their "times and seasons." We can hardly imagine it possible that little more than a century ago the "great Cham" of English literature declared that natural history was a study only fit for children! And we are thankful that we have grown to this—to regard the great life-scheme of our planet, past and present, including objects as minute as others are huge, and as structurally simple as others are complex, as one in its nature, evolved through the omniscience of an All-wise Being! If nothing less than Omnipotence could have produced it, surely we cannot but esteem it one of the noblest studies in which the human mind can be engaged. Science is one with the Psalmist in regarding the inorganic and organic kingdoms of nature as doing His will—beasts and all cattle, worms and feathered fowls, mountains and hills, fruitful trees and all cedars, fire and hail, snow and vapour and stormy wind fulfil His word!

CHAPTER II.

THE HISTORY OF AQUARIA—*continued.*

THE most natural of all the artificial conditions under which fish were kept in readiness for the table were the old fish-ponds. Many of the latter were covered with the usual aquatic vegetation, which thus kept the water pure ; or else a stream regularly passed through the lattice-work at either end. The relation which plants and animals bear to each other was not *fully* known even half a century ago. It required considerable progress in chemistry before the gases which they gave off were understood. Unquestionably the first step in this direction was made by Dr. Priestley, of Birmingham, who observed that oxygen gas was given off by plants when under the active stimulancy of sunlight. Aquatic animals had been described by Trembley, Baker, Leuwenhoek, Hooke, and others ; but they either obtained them direct, or else, as Trembley did his *Hydras*, kept them in jars by constantly changing the water. Naturalists were not aware of the needlessness of their labour until a long period afterwards. Even Sir John Dalyell, whose minute investigations into the structures and habits of zoophytes were published in his splendidly illustrated work on 'The Powers of the Creator displayed in the Creation,' and in his 'Rare and Remarkable Animals of Scotland,' and who kept many of the animals alive whilst he was observing on or experimenting with them, did so by constantly changing the sea water in which they were kept. One of the most wonderful things in a modern aquarium, to a person ignorant of natural history, is that the sea and fresh water never want changing. Such people have not yet learned that the dry land of the entire globe is only one huge *vivarium*, and that the Atlantic and Pacific oceans, as well as all rivers and lakes, are likewise only immense natural aquaria. This well-being of terrestial and aquatic plants and animals is kept up and perpetuated without changing either the air or water. What naturalists strive after is, to represent these natural conditions as much as possible.

One of the first notices we have of the establishment of aquaria on the modern basis of adjusting animal and vegetable life, is that of Bordeaux, commenced by M. de Moulins, in 1830. This naturalist found that by keeping plants in the water where his fish and mollusca were, the latter were stronger and healthier for it. But the question shortly afterwards assumed a thoroughly scientific foundation, although the cautious way in which conclusions which now seem to us self-evident, were approached, may appear ludicrous. They were accepted, however, in the true spirit of scientific research, which ought not to take

anything for granted that has not been amply proved.
At the meeting of the British Association at Cambridge, in 1833, Dr. Daubeny showed that plants
when in water (and aquatic species particularly) gave
out oxygen and absorbed carbon under the influence
of light. After detailing his experiments he expressed
his opinion—an opinion which has since then not only
been proved true, but which is universally accepted—
that " he saw no reason to doubt that the influence of
the vegetable might serve as a complete compensation
for that of the animal kingdom." An old proverb
says : "A child on a giant's shoulders sees farther than
the giant." It is only the superficial who smile at
the strenuous efforts of great intellects to attain unto
a knowledge of those principles which we now regard
as self-evident and incapable of contradiction. There
is an evolution of knowledge as there is of animal
and vegetable life. Daubeny saw dimly less than
half a century ago what every teacher in physical
geography now imparts to his class—that the oxygen
generated in the virgin forests of the Amazons valley
may be brought by the wind to bring health to
the fetid streets and alleys of crowded European
cities, and that in return the carbonic acid breathed
forth from our over-populated towns may be carried
on the " wings of the wind," to be eventually absorbed
by the incalculable stomata which crowd the under
surfaces of the leaves in the same forest-clad region !

The labours of an unassuming but true naturalist,

Dr. N. B. Ward, did much towards proving to students of nature that the magnificent views of Priestley, Daubeny, and others were both true and capable of being practically applied. Mr. Ward, in 1842, published a little work which gave a series of experiments showing that animals and plants might be kept in air-tight glass cases, and that each might be so adjusted as to breathe in what the other breathed out. He had commenced this study in 1837, and the celebrated "Wardian cases" for ferns, now to be seen in most drawing rooms, are the popular results. Dr. Johnston, the well-known writer on 'British Zoophytes,' adopted the above-mentioned compensatory principle in 1842, at which time he had a store of sponges, zoophytes, &c., in course of artificial preservation for scientific purposes. These animals were kept in small vessels wherein had been placed the common *Corallina*, the sea lettuce (*Ulva*), and several others; and the result was so successful that he suggested the possibility of marine aquaria on a more extended scale.*

The knowledge thus gained by a few experiments was destined shortly to receive considerable accretions. In 1850, Mr. R. Warington (whose name is inseparably associated with the history of aquaria) made a communication to the Chemical Society on

---

* The first attempt to keep the sea water constantly fresh by the presence of living seaweeds was successfully carried out by Mrs. Anna Thynne, in 1846.

his own experience in keeping a fresh-water aquarium. It was of a very simple and unpretending character, and differed little from that to which Pepys refers in his 'Diary,' consisting merely of a glass globe of fresh water in which two goldfishes had been placed, together with some plants of *Valisneria.* The latter is one of the best oxygen-producers of all known aquatic plants, and has long been a favourite with aquarium keepers. By-and-by, Mr. Warington introduced some pond snails to eat away the green algæ which formed along the inner surface of the glass. Two years afterwards, he and Mr. Gosse experimented after a similar fashion with sea water. This was the commencement of that rage for small marine aquaria which shortly afterwards set in. Tanks were constructed for the purpose, and marine animals and plants introduced in such proportions as were hoped to neutralise each other's respiration.

The most marked epoch in the history of the marine aquarium, however, undoubtedly took place when Mr. Philip Henry Gosse's most charming books made their appearance. Their attractive style of description of the lovely objects which are to be found in the commonest rock-pools of our coasts, and which it is possible to preserve to constantly delight the eye, induced hundreds of people to commence aquarium keeping. Never before had the common objects of the seaside found a historian at once so charming and so accurate. And although,

after a time, a great many people took to some other
new "hobby," and allowed their aquaria to fall into
neglect, sufficient enthusiasm was created to keep up
the practice to the present time. Some of our public
museums, notably that of Liverpool, shortly afterwards exhibited small tanks or glass vessels, containing aquatic animals and plants so arranged as to
keep up an equilibrium. Mr. Gosse first began with
sea anemones, the easiest of all marine objects to
obtain and afterwards to keep in healthy order. A
collection of these, and of some scarcely less attractive
sea worms which he had made at Ilfracombe, were
purchased by the Zoological Society of London, and
transferred to the new fish-house which had just been
built in the Zoological Gardens. In making a further
collection for the aquarium which was opened there
in 1853, Mr. Gosse gathered most of the material that
shortly afterwards appeared in his work on the 'Marine Aquarium' and 'Rambles of a Naturalist on the
Devonshire Coasts.' The small aquarium opened in
the Zoological Gardens, London, in 1853, was the first
public one started in England, and, although it has
long been superseded, it has done good work. In the
same year another public aquarium was opened for a
short time at the Surrey Zoological Gardens. That at
Dublin, which commenced about the same time, was
more long-lived, and was remarkable for the ingenious
way in which the curator, Dr. Ball, supplied the tanks
with fresh air. He so constructed air-bellows that

the visitors to the aquarium worked them with their hands, as a sort of amusement in the intervals of pacing about examining the tanks, and the Doctor found the air supply thus administered was sufficient. It is, however, too uncertain a method for other institutions to copy.

In fitting up small marine aquaria, the chief difficulty which people found who lived inland was in getting good sea-water. To meet this want, in 1854, Mr. Gosse showed how artificial sea-water could be manufactured, by simply adding salts to pure fresh water. The now largely used artificial sea-baths are produced by a small modification of Mr. Gosse's recipe. So successful was the experiment that even great public marine aquaria, like those soon after-wards founded at Hanover and Berlin, were supplied with salt water manufactured after Gosse's fashion. As soon as it was found that no great labour was needed to keep marine and fresh-water animals alive and healthy, in simply aerating the water, or in having properly adjusted aquatic plants, public aquaria were commenced in many of the large towns in Europe. Although these were not of the pretentious character with which we have now learned to associate the name, they did much to develop an interest in natural history. Before long there were aquaria at Belfast, Galway, Edinburgh, Scarborough, Weymouth, Boston, Vienna, Hamburg, Cologne, and especially at Havre. Some of them consisted of only one huge tank, wherein the animals obtained fresh air either by pumping it in, or by the natural aeration of plants. If the tanks were large, however, it was found that the latter system was attended with a good deal of difficulty. Hence the large tanks were usually aerated by causing the water to circulate and be injected in sprays, or else jets of air were forced into the water, which thus came into contact with oxygen, at the same time giving up its carbonic acid. The interior of the large tank in the Havre Aquarium was fitted up with rockwork, so as to resemble Fingal's Cave, in Staffa. Similar devices, all of them in bad taste, have been adopted at the Brussels, Hanover, Boulogne, Berlin, and Cologne aquaria.

The first of those large public aquaria, which have lately grown to such colossal proportions, was that opened by the French Acclimatisation Society, in the Bois de Boulogne, in 1861. Its length is 150 feet, and it is fitted up with fourteen tanks, each of which contains two hundred gallons of water. Ten tanks are devoted to fresh-water objects, and four to marine. The aquarium at Hamburg, opened in 1864, has also been very successful. It has long been considered one of the best on the Continent; much of its success depending upon the fact that Mr. William A. Lloyd was the deviser, and for some time the curator. Under his able management the zoological department attracted a good deal of attention among naturalists. In Great Britain we have hitherto been

C

very fortunate in having men who are well known as
naturalists at the head of our large aquaria. Thus
Mr. Henry Lee has charge of that at Brighton. Mr.
W. Saville-Kent was for some time curator of the
Manchester Aquarium ; afterwards he partly superintended the erection of one at Yarmouth, and now
he is consulting naturalist to the extensive aquarium
recently built at Westminster. Mr. W. A. Lloyd
has had charge of the well-known Crystal Palace
Aquarium since its opening in 1871 ; whilst at Southport there is a careful and diligent naturalist super-
intending the aquarium in Mr. C. L. Jackson.

We may regard the establishment of the Crystal
Palace Aquarium as an important epoch in the history
of the great public aquaria in this country. Its success
undoubtedly stimulated that at Brighton into existence, and the fact that the latter paid a good dividend
(always an important one to Englishmen) was quite
sufficient to induce companies to start those at Manchester, Southport, Yarmouth, and elsewhere. The
size of the Crystal Palace Aquarium is 400 feet long
by 70 feet broad, whilst the frontage of the tanks
amounts to 390 feet. There are sixty large tanks
exhibited, besides those held as reserve. These contain 20,000 gallons of sea water, whilst there is a
large storage reservoir which holds 100,000 gallons
more. The largest of these tanks is 20 feet in length,
and holds 4000 gallons of sea water. The animals
within the large tanks are viewed through the glass

fronts. There are two adjacent rooms, however, in which stand twenty other tanks, varying in capacity from 40 to 270 gallons, where the animals are viewed from above, looking down into the water, as well as laterally. The number of fish, zoophytes, annelides, &c., kept alive in this splendid aquarium is very great, the sea anemones alone amounting to several thousand. Every one of the latter has to be fed separately by means of wooden forceps.

The Brighton Aquarium is the largest yet constructed in England, and its interior is perhaps the most ornately fitted up, and varied with natural objects. The chief corridor (that which contains the aquarium proper) extends 220 feet. The tanks are placed on each side. They are of various sizes, the largest being more than 100 feet long by 40 feet in width, and holds 110,000 gallons of sea water, or nearly as much as that of all the tanks and storage reservoir, included, of the Crystal Palace Aquarium. Indeed, this huge tank is big enough for the evolutions of porpoises, full-grown sturgeons, sharks, sea-lions, turtles, and other huge marine animals. The next largest tank is 50 feet long by 30 feet broad. This is placed immediately opposite the former. The total quantity of sea water contained in all the Brighton tanks is over 300,000 gallons, besides which there are storage reservoirs into which the salt water is pumped directly from the sea outside, which are capable of holding half a million

gallons more. The salt water thus obtained, however, is liable to be very turbid. This huge quantity
takes about ten hours to be pumped in. In the chief
corridor above mentioned the number of tanks is
twenty-one. The total frontage of all is about 740
feet. Octagonal table tanks are also exhibited, in
which the rarer marine zoophytes, &c., are kept,
and where the process of fish-hatching may be seen
going on.

The most important event which has taken place
in the history of aquaria, from a purely scientific
point of view, was undoubtedly the founding of the
aquarium at Naples by Dr. Dohrn, a German
naturalist, Mr. Lloyd aiding in its construction. The
expense was borne almost entirely by himself and
a few personal friends, but the result has been
scientifically successful. Dr. Dohrn's idea was to
make it a kind of zoological station for the observation of the life-histories of marine animals
analogous to astronomical observatories or stations.
The ground floor of the building covers 8000 feet,
there being a story above fitted up as a zoological
dissecting room and laboratory, for the use of naturalists. Further, Dr. Dohrn here receives students
of natural science, the animals examined being obtained by dredging expeditions which are carried on
from time to time. A certain number of students'
"tables" were offered to various Government scientific
societies at a fixed sum. Some of these were taken

by the Universities of Oxford and Cambridge for the use of students who might gain the right of study. The aquarium is fitted up with the usual tanks, &c., on the ground floor, and is opened to the public at a certain charge. The money thus received is applied towards defraying the expenses of the institution. Already some highly important natural history work has been done here, notably researches in the embryology of certain fishes.

Shortly after the Naples station was founded, a similar aquarium was commenced at Penekese Island, the expense of which was defrayed by the munificent act of one of the New York merchant-princes. It was placed under the charge of Professor Agassiz, who unfortunately died almost before the institution had got into working order. The undertaking is now under one of the Professor's sons, and the scientific investigations promised to be of great service to zoology, but it has not hitherto proved so successful as was expected. There is no reason in the world why all our great public aquaria should not prove as effective to pure scientific research as they already are to the public educationally. Practical students might be attached to each, whose time could be devoted to zoological research. The time of the curator, however scientific may be his attainments, must necessarily be too much taken up by the general management for him to carry out observations which require constant and assiduous watching.

The Manchester Aquarium was the largest inland institution of the kind, before that at Westminster was built. It was opened to the public in 1874, and for a long time was under the direction of Mr. Saville-Kent, F. L. S. The main portion of the building occupies a superficial rectangular area of 150 feet in length, by 72 in breadth. At each extremity of the saloon are placed the two largest tanks. These occupy the entire width of the room 40 feet; so that they are capacious enough to contain living animals of considerable size. The total number of tanks at present existing is sixty-eight. These have a linear frontage of nearly 700 feet, which approaches very nearly the total frontage of the Brighton tanks. It is contemplated adding a series of tanks between the arches separating the saloon from the corridors so as to raise the total number to one hundred. This would give an additional frontage of 224 feet, and so far would render the Manchester Aquarium the most extensive in this respect. In addition to the above, there is a number of octagonal table tanks for fresh water and the smaller and rarer marine animals.

The Southport Aquarium was opened the same year as the latter. It is well situated in the town, which may be regarded as the "Brighton" of the Lancashire coast. The climate here is milder than anywhere in Lancashire, so that it is a place to which invalids resort in the winter—hence the " Winter

Garden " which is associated with the aquarium. The tanks are constructed much on the same plan as those at Brighton, and have a total linear frontage of 500 feet. Another aquarium at Blackpool, an adjacent town on the same coast as Southport, has tanks possessing a frontage of 250 feet.

Other aquaria are in course of erection at Scarborough, Yarmouth, and elsewhere. That at Yarmouth is intended to have show tanks in which 200,000 gallons of sea water will be held. The building is now nearly completed, and is expected to be opened to the public during the present year. The extensive aquarium at Westminster is in connection with a "winter garden." Although opened to the public the tanks are not yet fully stocked. The show tanks will hold 150,000 gallons of water, whilst there are storage reservoirs underneath capable of holding 600,000 gallons more. Public aquaria are further either being built or contemplated at Rhyl, Rothesay, Plymouth, Torquay, Southsea, Tynemouth, Margate, Scarborough, Ipswich, and elsewhere ; and there cannot be a doubt that within the next few years, most of our large seaside, if not inland, towns will possess these useful and attractive institutions.

CHAPTER III.

PRINCIPLES OF THE AQUARIUM.

THERE can be no question that portable fresh-water and marine aquaria may become sources of endless amusement and instruction ; and at the same time be so constructed as to ornament the rooms in which they are placed. Fresh-water aquaria especially, may be arranged so as to add to the usual cheerful aspect of our English homes. The sight of the moving objects, and of the green water-plants covering and shooting above the surface of the water is undoubtedly cheering. Invalids, or people of sedentary habits, who are much confined within doors, might find comfort and enjoyment from keeping an aquarium. The antics of its little inhabitants, and the little care required to keep this miniature world in a healthy condition, will draw off their attention from many an hour of suffering or care, and unconsciously develop a love for God's creatures. To children, aquarium keeping may be the means of imperceptibly teaching those feelings of humanity towards the lower animals which have hitherto been too much neglected. The "hunting instinct" is strong in most boys, and a love of natural history might direct this so that it would benefit man and beast alike : not unfrequently it assumes the character of unconscious cruelty, and the possession of *might* soon passes into the belief that its exertion is *right* The thoughtlessness with which children often torture flies, worms, &c., must undoubtedly be the means of partially developing a nature that ultimately finds a pleasure in inflicting pain, or in causing death. There is only too much truth in the sarcastic remark that when an Englishman is on a visit to the country and writes home to say he is enjoying himself, you may be sure he is killing something ! Anything which can neutralise this tendency to cruelty, or develop a more tender regard for the lower organised of our fellow creatures, becomes a means of moral education. This, we contend, might easily be brought about by keeping an aquarium, and interesting children in the funny ways of its inhabitants.

Many people think a fresh-water aquarium "only gives a lot of trouble, and is always getting out of order!" Of course, there is no denying that both these conditions may easily be brought about; but cannot the same excuse be made for declining anything else ? An aquarium properly constructed, and peopled with proper inhabitants, gives very little trouble indeed. A few minutes now and then are quite sufficient to keep it in that active, healthy order which gives so much pleasure to the possessor. A little common sense exerted in its arrangement cannot fail to ensure the perpetual comfort of its inhabitants. There are few "hobbies" which require less trouble, and as a rule it will be found that whatever "trouble" is caused, is due to ignorance, in not understanding the habits of aquatic animals and plants and the conditions under which they best thrive.

But, it may be asked, how are we to know all about such matters, unless we gain experience by first keeping aquaria? This is very true, but unfortunately a great many people do not persevere in keeping them, but exchange them for some other amusement as soon as difficulties arise. Perhaps they have not understood the elementary principles on which a streamless aquarium should be constructed, and so in putting it together wrongly they have been laying up for themselves an endless store of trouble. Or they have not taken the slightest pains to understand whether the animals and plants they have placed in the water were likely to agree there. An aquarium thus stocked has looked well up to the evening of the same day, but next morning it has presented all the appearance of a field of battle. In the darkness of the night, or rather in the early morning, a dire conflict has taken place. Each animal has been battling with its fellow, the weakest has gone to the wall, and only a few gorged cannibals remain of the too large stock with which the aquarium was peopled the day before! These are eyeing each other with suspicious anger, and it is evident at a glance that the war will be waged to the death.

We cannot too distinctly remember that a streamless aquarium is a little world, shut off, as it were, from the great world outside. The water, the animals, and the plants have to be so adjusted that no extraneous addition is required. The marvellous principles of adjustment of animal to vegetable life, and contrariwise, which holds good all over the surface of the globe, is as much in active operation in a portable aquarium as on a planet. Under the influence of sunlight the aquatic plants obtained from some stream or pond give off oxygen. You may frequently see it, in little bubbles, clinging to the stems and under surfaces of the leaves. We need not say that this gas is vitally necessary for the support of animal life. Plants therefore provide it. On the other hand, it is equally important that the carbonic acid given out by all animals shall be disposed of, or put out of the way so as not to injure the creatures that have breathed it, after the fashion of the Black Hole at Calcutta. Plants perform this function; and not only do they absorb the deleterious gas, but they actually require it for their sustenance and growth, as much as animals do the oxygen!

It will be seen, therefore, that in the knowledge of this fact we have the means of adjusting a collection of aquatic animals and plants, in the vessel we call an aquarium, so that there shall be constantly kept up a mutual compensation. The next important thing is to know how many animals we can place in a tank where there is already a certain number of plants. Unfortunately, people who commence keeping aquaria are usually too anxious to have as many and as varied a stock of animals as possible, and most of the evil which overtakes their endeavour arises from such over-stocking. It is evident that if there are more animals in the aquarium than there are plants to provide oxygen for, all of them will have to go short. This means universal sickness, and that pitiful gasping for air which is often to be seen in over-stocked aquaria. Before long it ends in death. Perhaps one or two of the weaklier die first. Their bodies lie on the bottom and are not removed. Decomposition sets in, and the water becomes fouler than ever. A white fungus—or rather the first stage of growth in many microscopic fungi—covers the bodies of the survivors. The aquarium becomes a painful scene of misery, disease, and death—a too vivid picture of similar conditions among humanity when the latter is horded in fetid and over-populated alleys, short of air, short of food, short of fresh, pure water! What wonder that many an enthusiastic young naturalist has been so thoroughly depressed by his first mistake terminating so fatally, that he has cast away the contents of his first aquarium, and never tried afterwards !

And yet all this evil has been wrought for want of a

little consideration, and perhaps because people could not resist the temptation to have as many animals in the aquarium as possible. There is one sure rule of guidance to a beginner in these matters—have too few animals rather than too many. They will compensate for numbers in the sense of health they seem to enjoy, in their vivacious gambols, and sprightly habits. The fishes are here, there, and everywhere, instead of always gasping with open mouths on the surface of the water.* The newts are frolicking about, or basking on the surfaces of the leaves and stones. Still, although we give this advice, it is to be followed with a degree of caution, for the over-stocking of an aquarium with plants is liable to overthrow the balance of life with almost equally fatal results. If there are too many plants the principle of natural selection soon sets in ; the weakly or badly adjusted species die off; the water becomes foul, and perhaps assumes a thick green hue. There is a ready means of checking such a disaster, however, for the evil resulting to an aquarium from excess of plant growth is not so rapid in its effects as when it is over-crowded

---

* Mr. W. A. Lloyd has produced twelve practical articles on Aquaria in the following numbers of 'Cassell's Popular Recreator,' published in 1873 and 1874 :—4, 8, 12, 16, 20, 24, 30, 32, 35, 38, 41, and 45. They are illustrated by fourteen woodcuts, of which eight are especially valuable as representing, drawn to an accurate scale, how many creatures, and of what kinds and sizes, can be maintained in aquaria of a named water capacity, of a given water distribution as to surface, and at a given temperature.

with animals. Moreover, when the first symptoms
set in you can neutralise them by adding another
animal—a small fish, a young newt, a few tadpoles,
or two or three water snails, adding them one by
one, and waiting to see the results. In this manner
you proceed as a chemist does when weighing some
valuable or important medicine. He trickles a little
at a time until he has attained the exact weight to a
feather. Excessive growth of plants may be kept
down in one or two ways. First by *modifying the
light.* Owing to the strong desire to see as much as
possible of what is going on in the aquarium, many
people expose it as much to the light as they can.
And, as they have perhaps fallen into the other
mistake of constructing three, if not four, of the sides
of glass, it follows that the amount of extra stimulus
to which the plants are exposed far exceeds that
which influences them in a state of nature. In a pond
all the sides are dark—the light can only get into the
water from above, or through the surface. In a river
or stream the sides are always dark, and, though the
light can reach the bottom from behind and in front,
it passes wholly from overhead. When there is too
much glass used in the construction of an aquarium
there is a temptation to use the glass : this means
exposing the aquarium to light, so that the latter
passes completely through it on every side. For this
reason *bell-glasses* are specially to be shunned for
fresh-water aquaria. The young beginner has only to

remember that the secret of his successful preservation
of animals and plants lies in his imitating natural
conditions as much as possible. And a very slight
consideration will show him that round bell-glasses,
and square tanks, having three or four glass sides, are
as far removed from these as they well can be, unless
in shady places.

It is this intensity of light which promotes the
rapid growth of the greenish film coating the inner
surfaces of the very glass through which you want to
watch your animals, as if it were a judgment inflicted
on your unscientific taste ! This is not the worst of
it : the same lowly-organised and rapidly-developing
algæ will mantle your water plants with their green
slime, and strangle and suffocate them in folds of
sickly-looking greenery. At the slightest sign of
anything of this kind occurring, the aquarium ought
to be put away from the light. A few days in a
darker corner will soon restore it to its healthy conditions, if only the disease has been taken in time.
Another means of keeping down the excessive growth
of aquatic vegetation is by introducing more marsh
snails, such as *Paludina*, the larger species of *Planorbis* (*P. corneus*), &c. These crawl over the inner
surfaces of the glass and clean it, removing and devouring the green film ; they also keep down the
tendency to too rapid growth in *Anacharis, Callitriche,*
&c., on account of their fondness for the young and
growing shoots.

Another evil threatening all aquaria results from not attending to the *temperature* of the water. You see tanks placed full in the sunlight of the window, where the inhabitants are most exposed to the light and heat. We have noticed how the light thus received encourages the undue growth of vegetation ; now we have only to remark on the sickly condition of the water when it is so thoroughly and directly heated by the sun. The temperature is raised far beyond what it could be in a pond or a stream. In the former the sides and bottom are always dark and cool, and in the latter the greater ease with which the sun can heat the shallow water is compensated by the constant change of the water in running. In an aquarium placed in the full sunshine it is evident that the animals and plants alike are exposed to most unnatural conditions. With the elevation of the temperature there is attended a less capacity for the water to contain the mechanically-mixed oxygen given off by the plants, or even to absorb it from contact with the air at the surface. Animal matter decomposes more readily, and thus the water becomes sooner fetid. What the student ought especially to observe, therefore, is that his fresh-water aquarium is placed where the temperature of the water *varies* as little as possible. It ought never to fall below 40° or rise higher than about 60°, if he desires to keep a healthy stock of animals and plants.

A well-constructed aquarium ought to continue

in the same state for years. It is a common error among those who have had no experience in these matters (and very likely the notion has fostered the idea as to the great trouble which aquaria give) *that the water ought often to be changed.* Nothing of the kind. A well-balanced aquarium, one that has eventually "got into good working order," wants no water added to it except *what may be lost by evaporation.* If proper care be taken, even this may be reduced to a minimum. The best plan is to have the top covered with a plate of glass. It may be loosely placed there, and ought never to be fastened down, else there would be no means of getting at the contents of the aquarium. Such a plate of glass lessens the evaporation, and protects the surface of the water from dust. If you desire that the aquarium should be further ornamental, this may easily be done (with one of sufficient capacity), by having a fountain playing in the middle. Fewer aquatic plants are then required to aerate the water, as the fountain does it mechanically, entangling films of air on the surface of every drop of water thrown up. The plants may then be ornamental, such as the water violets (*Hottonia palustris*), water plantains (*Alisma plantago*), &c. All that is required is a wide-mouthed bottle, in the cork of which are three holes though which the glass tubes seen in the sketch pass. c reaches nearly to the bottom, whilst the other two pass only through the cork. A is a wide, funnel-topped tube. c is bent at

the top, and has there attached to it a piece of long
indiarubber tubing.   The cork and tubes should fit
perfectly.   In order to set this easily improvised

Fig. 1.

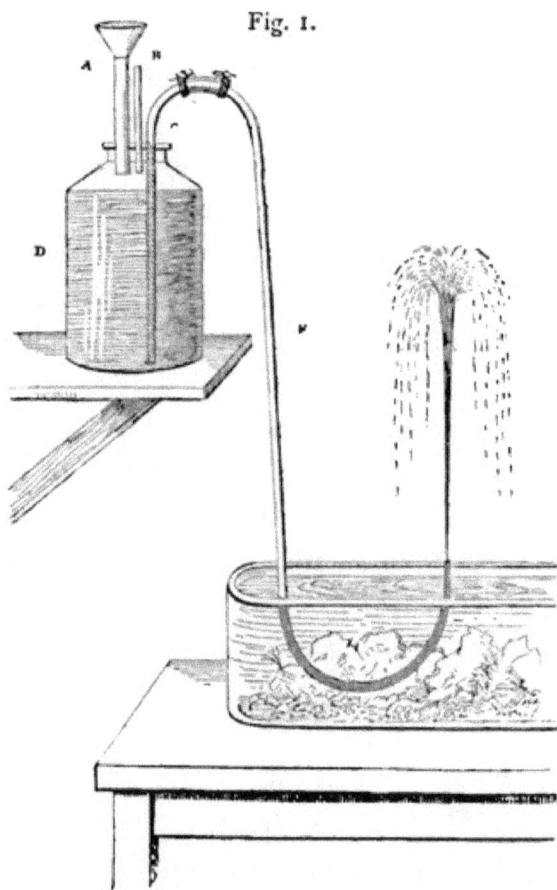

Extemporised Fountain for small Fresh-water or Marine Aquaria.

fountain in action you fill the bottle.   When it is full,
continue pouring water gently into the funnel until
it is above the level of the bend in the tube c.   A
little will then flow over into the long leg of the

syphon E. The water will of course continue to flow until the level of the water in the bottle falls below the mouth of C. The tube B is for the escape of air whilst filling. A very short experience will enable the student to work this cheap fountain, and it is evident it will flow for a greater or less space of time according to the magnitude of the feeding bottle and the bore of the indiarubber pipe, which is bent upwards at its extremity for the purpose of throwing the water into the air. It is true, more water is wasted by evaporation in this manner, but this is a difficulty easily met, as sufficient fresh water can always be put in the service or feeding bottle. The fresh-water aquarium may be made prettier and more ornamental than it hitherto has been, with aquatic flowering plants, if only pains be taken to render their conditions of growth natural. There is no reason why we should not have aquatic gardens of this kind in our rooms.

The dust which accumulates when the presence of such plants renders a closely-fitting glass plate impossible, can easily be removed now and then by gently laying pieces of blotting paper on the surface of the water. The dust adheres to it, and it is then easily removed. Or it can be skimmed off by using the edge of a sheet of writing paper. By a little skill and care, we might easily possess semi-aquatic gardens in which miniature fountains are made to play; and the whole rendered a fit and healthy habitat for such creatures as can best be supported.

CHAPTER IV.

CONSTRUCTION OF FRESH-WATER AQUARIA.

THE construction of a moderate sized, portable fresh-water aquarium may be as cheap or as expensive a matter as a person thinks fit, or his pocket can afford. They can usually be purchased at the natural history dealers' shops in London and elsewhere ; but perhaps one learns more of the conditions under which the animals we propose to keep will hereafter live, if we have the aquaria constructed under our own super-intendence. Having fully taken into consideration the principles which ought to guide us in maintaining aquatic animals and plants, the next thing is to be sure the aquarium will not leak ; and that it contains nothing in the materials composing it which are at all likely to be poisonous. Under the direction of a car-penter or plumber, any of the aquaria of which we give illustrations may be constructed. One of the cheapest, perhaps, is that shown in Fig. 2, and, by a little alter-ation in the internal details of rockwork, &c., it may be used for marine or fresh-water objects as the owner thinks fit. The back and sides are composed of strong, half-inch wood, dovetailed together. The bottom is thicker, and is screwed to the framework. The front only is occupied with plate glass, which is let in by a kind of "rabbit and bead," as carpenters call it. The whole of the interior of the woodwork, back, bottom,

Fig. 2.

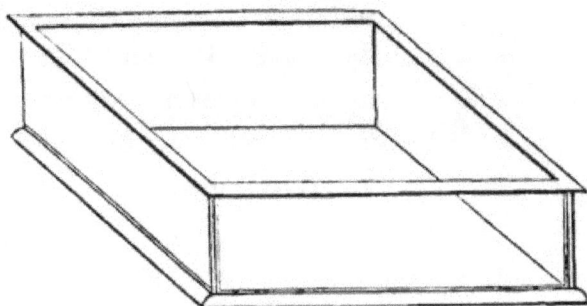

Cheap Portable Fresh-water or Marine Tank.

and sides, is then coated with pitch to the thickness of about one-eighth of an inch. Hot pitch is also run into the "rabbit," and the plate-glass front pressed well against and into it. If a wide beading is then run all round the top, the aquarium will be completed. Thus constructed, the whole expense will not exceed 14s. or 15s. Before stocking it with animals and plants, the tank should be seasoned in rain water for a week or two; and can then be used without any fear of leakage or harm. A costlier method of constructing a tank on the same pattern is to have the bottom, back, and sides of slate, instead of wood, with a plate-glass front as before.

There can be no doubt that aquaria with flat sides are much better than round bell-shaped glasses. They do not distort the objects when moving about, after the fashion in which goldfish often present themselves to our notice in the ordinary globes. For a few pond snails and a plant of *Myriophyllum*, &c., as "stock," an inverted bell-glass with a wooden base,

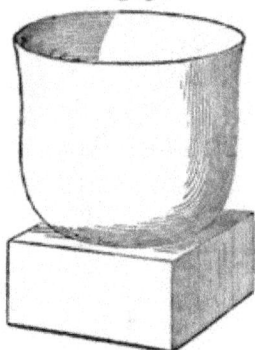

Fig. 3.

"Stock" Glass.

such as is shown in Fig. 3, may be used. It should be remembered, however, that only a very few objects can be thus accommodated; but if the owner have self-denial enough to forego the temptation of over-stocking the glass, such an aquarium may be healthily kept, and will even form a very pretty and lively little ornament to a room. Again, a darkened bell-glass may be used as part of a more elaborate attempt (Fig. 4), in which, by means of an ordinary cheap wire stand, it may occupy the centre and be surrounded with the ordinary flowering plants with which we are in the habit of decorating our rooms. A glass sheet protects the surface of the water in the aquarium from dust. The late Dr. Lankester, who was one of the best and earliest writers upon aquaria, showed in his 'Aquavivarium' that such an arrangement as this might be very easily and cheaply carried out.

Another inexpensive tank, which answers well for window purposes, providing the sides and back are made of opaque material and not of glass, is shown in Fig. 5. The top may be fashioned of wood or zinc slightly perforated, and should have a narrow plate of glass let into the top. If the front only be made of

Fig. 4.

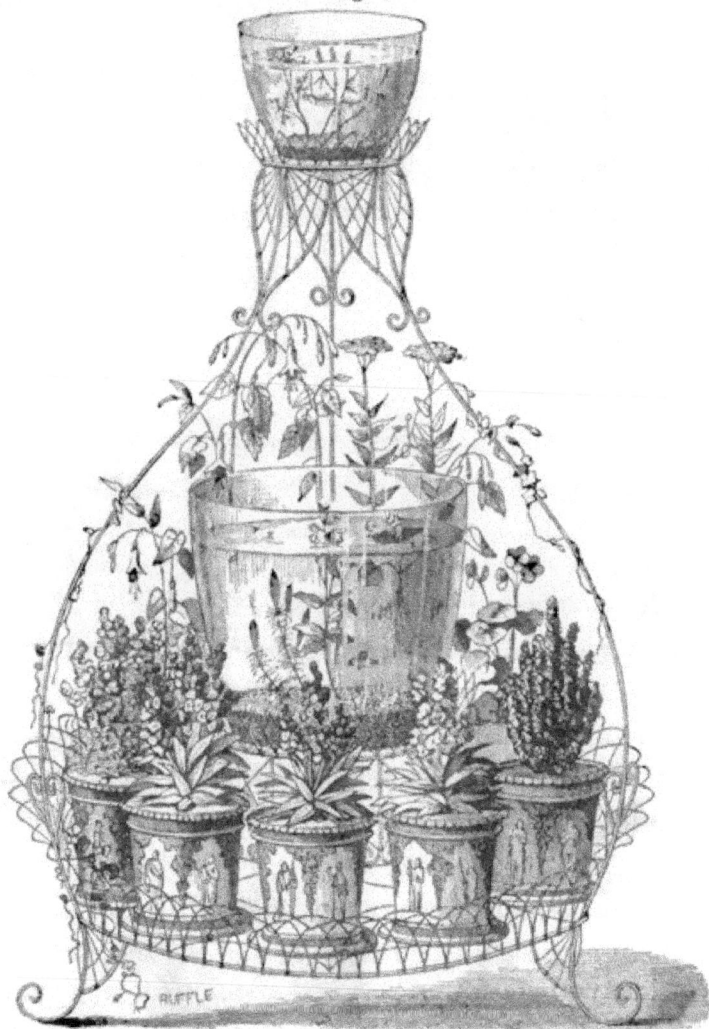

Flower-stand, with Bell-glass Aquaria.

glass, the light will not prove too strong. We have used this kind of tank both for fresh-water and marine

objects.   It should not be placed in a window having
a south front, as the light then is too strong, and will
develop that pest of aquaria, a thick opaque green-

Fig. 5.

Oblong Tank for Window Aquarium.

ness, do what we may.   A northern aspect is always
the best for the glass frontages of aquaria to face,
whether they be placed in windows or in rooms.

   In the construction of square or polygonal aquaria
it is of course necessary above all things that the sides
should be perfectly water-tight.   Leakage is a source
of annoyance and untidiness in a room, besides interfering with that balance of animal and vegetable
life which is sought to be sustained in the water.   The
following will be found a good receipt for making a
cement that will keep the sides and joints of a tank
perfectly water-tight: fine white sand, one part;
litharge, one part; resin, one-third part, mixed into a

paste with boiled linseed oil, and applied *unstintingly.*
A little rockwork is always an additional element of
attraction in an aquarium, especially if fish or amphibia are kept. In fresh-water aquaria, however, it is
never required to the extent it is in marine. When
built up loosely, the darker places afford a screen
from a too intense light, and those creatures which
cannot bear it will soon discover such retreats. But
this adaptation applies more to marine animals than
the fresh-water forms of which we are especially
treating. Newts love to crawl upon a stone or piece
of rockwork projecting above the water, in order to
bask in the sun; but this they will do if the water
plants are strong enough to bear them on the surface
of the water. The best and most harmless material
for rockwork is pieces of pumice-stone, fragments of
melted glass bottles, and such fragments of vitrified
bricks as may be picked up in any brick-kiln. These
should be fastened together with Portland cement.
In order to make the contents of the tank as light as
possible, one or two inverted flower-pots may be
fastened to the bottom. If the inverted edges be so
broken as to form a passage through, then the interior
will serve as a dark cave to any of the animals requiring such a retreat. The holes (now uppermost)
should not be filled up, for the water within the
inverted pot will be kept colder, and thus a healthy
current action between it and the surface water may
be set in action through the holes. Flower-pots thus

serve a double purpose. They render the rockwork built around and over their external surfaces (except where the edges are broken at the bottom to form a tunnel, and the usual opening in the middle of the inverted base) lighter than it would be if it were heaped up, one solid piece on another. And we have already seen that the colder bottom and warmer surface waters will set up a feeble vertical current action.

If aquatic plants are intended to form a part of the stock contents of the tank, the best plan is to procure them when young from some dyke or pond, and plant them in flower-pots. These flower-pots may be hidden among the rockwork ; nay, the latter may be loosely fastened around them by means of Portland cement so as to completely conceal them. Such species as the water violet, water plantain, water soldier, and arrow-head grow best when thus treated ; and as their flowering spikes ascend above the water whilst their leaves are mostly either submerged or floating, they form very pretty accessories to the larger fresh-water aquaria.* In the arrow-head, water ranunculus, and several others, the floating or surface leaves are of a different size and shape to the submerged leaves. All the plants just mentioned require a good depth of soil to be planted in, and their transference to flower-pots prevents

* Such aquatic plants ought not to be kept where gas is lighted at night, as they are then unduly forced, and pine away from not obtaining their necessary repose.

the necessity of filling the bottom of the tank with
the depth of soil or mud in which they require to grow ;
and as these vertically-growing plants might be placed
around the sides or at the ends of the tanks, more
room would then be left for the evolutions of fishes or
other aquatic animals. Hence, rockwork in the centre
of fresh-water aquaria is to be shunned, as it both inter-
feres with the movements of the objects and prevents
us witnessing them. The surface of the water should,
if possible, be partly covered with vegetation, for it
keeps the water cool and forms a retreat for the smaller
inhabitants, and also to some degree prevents undue
evaporation. One or two of the many species of
duckweed (*Lemna*) are useful in this respect, for none

Fig. 6.   Fig. 7.

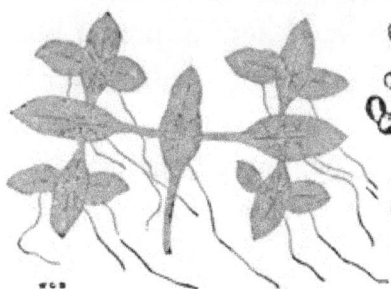

Ivy-leaved Duckweed (*Lemna trisulca*).
Lesser leaved Duckweed (*Lemna minor*).

of them need any soil. They derive what nutriment
they require from the water in which they float, sucking it by means of the slender rootlets which may be
seen let down from the layer of green frond-like

leaves. These roots terminate in a spongy base, through which the work of absorption is carried on.

In forming aquaria wherein it is intended to grow aquatic plants, a good deal of attention ought to be paid to the fact whether such plants require much or little soil. With the exception of the duckweeds, all require some, if only to anchor their roots in. But this mud should be dispensed with as much as possible, on account of the tendency there is to thicken the water whenever fishes or other animals stir it up. The great water beetle (*Dyticus*) very often does this, especially in the night time; and so you awaken some morning to find the water, which was clear the night before, in what seems a hopelessly muddy condition! Little if any soil is required by the *Anacharis* (a Canadian plant), one of the most useful an aquarium keeper can have if kept in proper bounds, for it grows only by shooting or budding, never by seeding, and many aquatic animals, fishes especially, are very fond of nibbling at the young green shoots. Another plant, equally useful and even more beautiful, is the star-wort (*Callitriche*), which requires a little sediment for its roots to be planted in. The leaves of the star-wort are much used by fishes and amphibia for depositing their spawn upon. The water crowfoot (*Ranunculus aquatilis*) requires little soil for its roots, but its needle-shaped leaves soon branch through and fill up the interior of the tank if too strong and old a plant be introduced. It is worth trying a little of this common plant, however, if only for the sake of the pretty kidney-shaped floating leaves and its brilliantly white and abundant flowers.

The best soil that can be selected for the purpose of covering the bottoms of fresh-water tanks is fine river sand, in which may be mixed a few small round stones. All should be well washed, or they may be the means of introducing into your aquarium objects you never bargained for, which will upset the equilibrium you so much desire to commence and maintain. A few pieces of charcoal mixed with it, are serviceable in absorbing all decaying organic substances. These, however, should be removed from time to time, when it is supposed they have taken up as much organic matter as they can. Charcoal also prevents foul smells, and generally acts as a deodoriser. Too much of it, however, is likely to be injurious. In fitting up an aquarium the soil should be placed after the rockwork has been constructed and the water plants rooted in their concealed flower-pots. To prevent the soil being washed up and the water rendered muddy, the water must be poured in through the finely perforated rose of an ordinary watering can. The fresh water commonly used for drinking purposes will do, and this is perhaps better than if obtained from a pond, where there is likely to be much more diffused vegetable matter. The water weeds, both potted and planted, ought also to be well washed before they are transferred, otherwise fish, amphibian, or molluscan spawn may be adhering to their stems and leaves, so that they are afterwards hatched.

Let us suppose that all these directions have been attended to, the plants are growing in the recently poured in water, and the latter is clear and transparent. Now let it remain in this state for a day or two, when the plants will have recovered from the shock occasioned by their transference, and the water will have been tolerably well charged with the oxygen they have given off. Then add one or two animals, a couple of minnows or newts, and a water snail or two, and watch the results. From what we have already said in the last chapter you will at once perceive whether there is a redundancy of animal or vegetable life, and be able to modify their relationship accordingly. When there is evidently a balance, endeavour to keep it. All the objects, animal and vegetable, which you have stocked the tank with, are mortal, and will sicken and die. You must at once remove dead bodies, or they will taint the water. A pair of wooden forceps will be handy for picking them up from the bottom where they are sure to be found occasionally lying. A little hand-net will also be found useful both for removing specimens and procuring them for further and minuter investigation.

If fish are kept they require a little food, but the quantity is so small that it does not do to provide it for them artificially. On no account get into the habit of feeding fish or newts with pieces of raw beef, or even earth worms. Whether they will accept them or not depends upon their capricious humour, and if they are not eaten they only accumulate on the bottom and foul the water. A much better plan is to keep as many snails as you possibly can. Their spawn is a favourite food with fish, and what with that, animalculæ, entomostraca, and the fresh shoots of aquatic plants, they manage to make all the feeding they require. If freshwater animals have to be artificially fed the best diet is the blood-worm (*Tubifex rivulorum*), which is to be found in many ponds and streams. Many fish eat it greedily, and it has the advantage of being an aquatic worm, so that if it be not eaten at the time you put it in the tank, it lives to be eaten another day! The fine blood-red colour of this worm is due to the false-blood being visible through the thin skin.

We have seen that the gradual change of the water to a greenish hue is due to the presence of minute vegetation which the undue light has stimulated into existence, and that the remedy for this was to subdue the light until the evil was overcome. Now we have to notice another unhealthy condition of things to which the inhabitants of a fresh-water tank are liable. Occasionally the weeds are covered with a white hairy slime, and the water gets thick with a muddiness which is evidently due to the same cause as the condition of the plants. Perhaps you will also notice the fishes gasping on the surface, as if they were unwilling to breathe the foul air mixed with the contaminated water. These symptoms *are due to want of air;* and when you perceive them, at once remove the aquarium to where it will receive an extra stimulancy from sunshine and oxygen. If this does

Fig. 8.

"Convalescent" Glass, containing *Hottonia*, &c. *Vallisneria* is the best restoring plant.

not immediately remedy the evil, remove one or two of the fish or newts, or other of the larger animals to a temporary glass, such as is shown in Fig. 8, where there are plants of *Vallisneria*, &c., growing. This is one of the best oxygen-giving, fresh-water plants we have, and may be used in small temporary jars or glasses for the purpose of so aerating the water that it acts as a restorative, and the glass thus serves as a kind of convalescent hospital. If a large aquarium is kept, or more than one, an infirmary of this kind will be found very useful.*

We repeat it, that the only successful way in which aquatic animals and plants can be maintained in a healthy condition, is by endeavouring as far as possible to imitate natural conditions. This, however, presumes

* It will be noted that all the foregoing directions apply only to *streamless* fresh-water aquaria. When opportunity affords (as it often may do) of constructing them so that the water may be conducted by a pipe from the usual household supply, and a constant circulation and aeration can be kept up, much labour will be saved, and the objects will appear more healthy and active.

a more intimate knowledge of natural history than
most people possess, and in such cases, therefore, all
we have to do is to attend to the above general in-
structions, until such time as experience will have
been able to suggest some other course. Every set
of animals and plants has different habits to another,
and we should never fall into the error of supposing
that because we have been successful with one group,
exactly the same treatment cannot fail to be effective
with another. To the aquarium keeper as well as
to the profound naturalist, the motto of Longfellow
equally applies—

"Learn to labour and to *wait*."

CHAPTER V.

AMPHIBIANS AND FISHES OF THE FRESH-WATER AQUARIUM.

IT needs little scientific knowledge to perceive that an aquarium keeper is likely to be far more successful if he attempt to keep a few animals, than if he over-crowd his tank with many.   It cannot be too strongly insisted upon, that more than half the misfortunes and so-called "bad luck" which are ordinarily experienced in the keeping of aquaria, are due to *over-stocking*. When a few objects only are kept it is surprising how healthy and vigorous they appear.   Moreover, they sooner get tame, or rather accustomed to their keeper, than when they are numerous.   It is always best to keep more than one individual of the same species if the tank be large enough, otherwise there is a sense of loneliness suggested which detracts from the pleasure of preserving animals ; and before long one sees that the solitary pets feel this themselves.   Two small fishes or newts are always preferable to one.

Nearly all our native species of animals can be thus kept in captivity.   Recently such amphibians as the pretty yellow-spotted salamanders, and those still more curious creatures the Mexican axolotls have been introduced, and aquarium keepers can now purchase them alive at any of the London naturalists. Still, we doubt whether either of these exhibit so much intelligence as our own newts, or if they exceed them in beauty. The great water newt (*Triton cristatus*), Fig. 9,

Fig. 9.

Great or common male of Water Newt (*Triton cristatus*).

notwithstanding the roughness of its warty skin, has a bright orange colour on the under part of the body which gives it a very attractive appearance. Its movements in the water are even more graceful than those of fishes. These animals have long been regarded with dislike and suspicion, and not many years ago farmers believed they could cause rheumatism and paralysis to cattle by creeping over their limbs. Even yet this superstition may be found lingering in out-of-the-way corners of England. We have ourselves heard mysterious diseases and complaints in cattle attributed to their drinking pond water in which newts were known to be abundant! The readiness with which country lads pelt newts to death even yet, is a "survival" of this ancient and ignorant prejudice. We need not say how thoroughly without foundation is this notion, or descant on the cruelty to which it has given rise.

E 2

The presence of a serrated crest along the back is characteristic of the males both of this species and that of the smooth newt. The latter, however, may easily be identified by its smooth skin and smaller size.

Both male and female of the great warty newt are easily tamed. We have kept them until they would come to the top of the water and take a worm from our fingers. They are voracious feeders, but it is best not to supply them with too much food. When they are in season, the tadpoles of the common frog will be found the best diet to give them. These do not taint the water as worms are apt to do when they die; and it is very interesting to witness the schemes and pursuits the newts indulge in to capture their second cousins. It will be as well not to keep this species and the smooth newt (*Lissotriton punctatus*) in the same tank, otherwise the latter may fall a victim to the ready appetite of the former. Even if it does not it is placed in hourly dread, and shelters itself so that it can rarely be seen. All the newts use their vertically flattened tails for swimming. The weakness of their legs on land adds to their reptilian appearance, for they are obliged to crawl, and are by no means so agile as the lizards for which they are frequently mistaken. On hot summer days you will see the newts, with their legs extended, floating and basking on the surface of the water.

It is only in the spring and summer months that

the males of these two species of newts have the dorsal crest fully developed. Like many other animals, birds especially, the sexes are reduced to a common likeness during the winter. When the warmth of later spring begins to be felt, it is astonishing how quickly the dorsal crest, and the characteristic colours and the spots of the males, are developed. In May and June these will be at their height, for the females are then depositing their ova, singly, in the folds of the leaves of the water plants. The eggs are soon hatched, and, as is well-known, the tadpoles are endowed with external gill-tufts (Fig. 10).

Fig. 10.

Tadpole of Newt (three months old).

The great warty newt is the best for the fresh-water aquarium, on account of its greater fondness for the water. It rarely leaves it, except to bask on the leaves, or on some stone. Hence it is as well to have a little rockwork projecting above the surface of the water in which these newts may be kept. During winter, it will lie torpidly at the bottom of the tank ; but if the latter be always kept in doors (as it ought to be), the period of hybernation will be very brief. Perhaps the reason why the crest is lost in winter is that it becomes absorbed, in lieu of food, by the system, to maintain the action of the involuntary organs. There is a popular error that the tadpoles of frogs and toads *drop* their tails and gills when they leave the water for the land. The real fact is that both these organs are absorbed and utilised, and are not dropped or shed at all.

The smooth newt is quite as common as the warty, and its habits are perhaps quite as interesting to the observer. The dorsal crest of the male is not toothed,

Fig. 11.

Adult male of Smooth Newt.

Fig. 12.

Adult female of Smooth Newt.

like that of the warty newt, although it is wavy in its outline. Both male and female indulge in graceful evolutions, and not unfrequently may be seen chasing each other in frolicsome sport. The female is exceedingly cautious in selecting the proper places for the deposition of her eggs ; and the process of laying them singly or in pairs, and afterwards of folding up the leaves of the plant around them, so as to screen them

from the keen eyes of other aquatic animals on the
look-out for food, is very interesting. If the plant be
present, the female smooth newt always seems to
prefer the leaves of the
Callitriche for this purpose.
This plant is very common,
and is one of the best that
can be selected for fresh-
water aquaria. The crea-
ture may be seen examining
one leaf after another until
she has selected one that
appears to answer her purpose better than the rest.
The eggs are laid at an
interval of three or four
weeks. Under the micro-
scope these eggs become
very interesting objects, inasmuch as the transparent
membrane allows every
stage of the development
to be plainly seen.

Fig. 13.

Callitriche, showing leaves folded
over eggs of Smooth Newt.

The claspers are used by
the young tadpole to hold
on to any object there is in the water, for it is only
after a brief experience that it is able to regulate and
control all its own movements.

Water fleas (*Daphnia*) and the blood-worms already

mentioned, are the best food for the smooth newts.
If these are supplied in abundance the newts will
rarely leave the tank.   Indeed, there is much reason

Fig. 14.   Fig. 15.

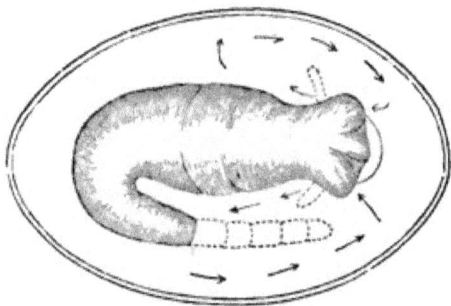

Tadpole in egg eight days after laying.
The arrows indicate the current motion
caused by the cilia. The dotted lines show
the increased growth eleven days after the
egg was laid.

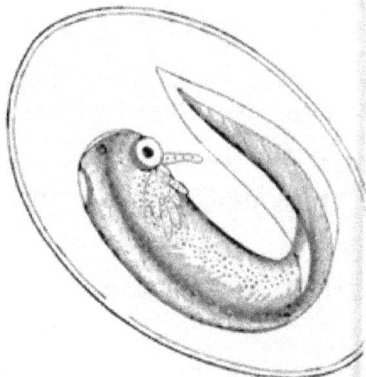

Development of tadpole in
seventeen days after the laying of
egg.

Fig. 16.

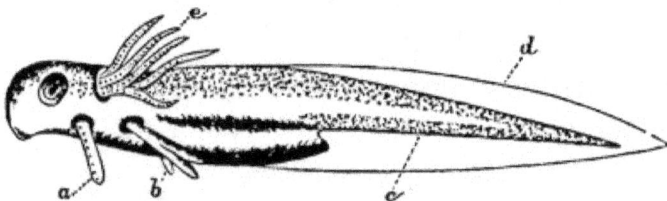

Earliest stage of the *free* tadpole of smooth newt—*a*, claspers ; *b*, fore-
leg partly developed ; *c*, circulation of the blood ; *d*, transparent fin ;
e, branchiate, or gill-tufts.

to suppose that one cause of their leaving is the absence or shortness of food supplies. They are very
curious animals, and will come to the inner surface of
the glass of the tank to examine anything unusual,

or even when you are observing the movements of any of the creatures through your magnifying glass. One of their habits is that of casting their skin, which is sloughed off whole, so that it can be afterwards collected and mounted. This process usually takes place when the breeding season is over, and male and female are assuming their winter skins.

We have one or two other native species of newt which are much rarer, and more locally distributed than the above, which would do equally well as objects for the aquarium. These are the straight-lipped warty newt (*Triton Bibronii*), and the palmated smooth newt (*Lissotriton palmipes*), found near Tooting. The development of the spawn of frogs and toads might also be usefully studied in an aquarium, although care would have to be taken that the quantity introduced was not too great for the aquatic vegetation to supply with oxygen. Very little attention has hitherto been paid to the development of these common objects, but there is no reason to believe it would be less interesting than that of the newts. Although resembling each other so much, the spawn of frogs and toads may easily be identified by the fact that the former occurs in lumps and the latter in single strings. The eggs of toads are about a quarter of an inch in diameter, and usually smaller than those of frogs.

Where there is an abundance of newt or frog spawn developing, several small fishes might be preserved to keep down its undue development, always providing that the capacity of the tank is sufficient to allow of necessary vegetable growth to provide them with air.

Of all the favourite species the goldfish has long been most domesticated, so that now, like the canary among birds, it seems to be better adapted to confinement than even to a free roving life. It will answer admirably in a tank supplied with *Anacharis*, the tender shoots of which it eats with great relish. This and an occasional blood-worm will serve for all the food it requires. Where fish are kept the utmost care should always be taken that no bread or biscuits are ever given to them, as these not only injure the fish, but contaminate the water by their speedy decomposition, unless the water is exceedingly well oxygenated. Care has also to be bestowed upon the *Anacharis* in such tanks as may contain it, for, as its only method of reproduction in this country is by budding, the latter process is apt to take place so vigorously as speedily to fill the water with a densely-crowded mass. It may always be made to grow by planting a sprig in a flower-pot containing soil, and placing this among the rockwork, so as to be hidden away.

Undoubtedly those of our native fishes which are easiest to obtain and domesticate are the sticklebacks and the minnows. We have several species of the former, some of which will live equally well in fresh and salt water aquaria ; and as some of them indulge in the unfishlike recreation of *nest-building*, they become really very interesting objects when the

tank is clear enough to enable one to witness their
habits of life. They are, moreover, pugnacious little
fellows, and will attack other aquatic animals, or get
up fights among themselves, as if they had been
geographically limited to the "Emerald Isle." You

Fig. 17.

Three-spined Stickleback engaged in nest-building.

may witness the male fishes carrying bits of weed, &c.,
in their mouths, and building up a heap in some corner
in this manner. This work finished, their next task
is to induce some of their female companions to come
there and deposit their eggs. Whilst engaged in this
courtly office the males assume the most lovely prismatic hues, especially about the head and shoulders.
The females deposit their eggs or spawn, and then
leave it to their male companions to defend the spot

against all spawn-loving fish, until such time as it is
hatched. Even then the labour is not over, for we
may see the paternal fish hovering about the young
fry with the greatest anxiety, and valiantly running
tilt against all other animals, those of their own species included, who may be desirous of making further
acquaintance with them. The commonest of these
little fish is the rough-tailed or three-spined stickleback (*Gasterosteus trachurus*). The male will attack

Fig. 18.

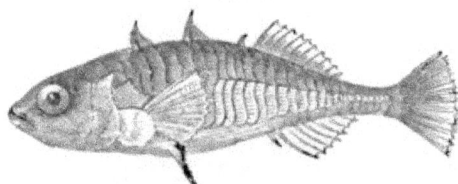

Rough-tailed Stickleback (*Gasterosteus trachurus*).

anything that comes near his nest, even the carnivorous
water tigers and water beetles. Another species is
Gasterosteus semi-armatus, very common in some
streams and rivers, and this also constructs a nest,
and generally adopts the habits above described.

Nearly allied to the sticklebacks in many respects
is the Miller's thumb, or river bull-head (*Cottus
gobio*). In many rivers they are very abundant, so
there is no difficulty in procuring them. In the
aquarium they are active and wary little animals,
loving to hide under the loose stones.

The loach (*Cobitis barbatula*) is another interesting
little fish that may be readily domesticated. As its

barbules would lead anyone to infer, it is a bottom
feeder, and its colour and markings are such as readily
to hide it from observation when in its natural habitat.

Fig. 19.

The Loach (*Cobitis barbatula*).

Like the miller's thumb, it often hides under the
stones, or in the interstices of rockwork, but it is by
no means so voracious. In the aquarium it, as well
as most fishes, is very useful in devouring any odd
worm or dead insect or fragment that may have been
lying on the bottom of the tank, and that otherwise
would have decomposed and contaminated the water.

The gudgeon (*Gobio fluviatilis*) is even a greater
favourite with some aquarium keepers than stickle-

Fig. 20.

The Gudgeon (*Gobio fluviatilis*).

backs. It grows to a larger size, often to 6 inches in
length, and is therefore a more attractive object. It
may readily be identified by the pair of barbules, or

feelers, one on each side the mouth. If this fish is kept, the bottom of the tank should be covered with sand and gravel, not mud, or otherwise the water will be fouled by its habit of stirring up the bottom. It is a voracious feeder, and will eat up any animal garbage that may be lying about. It is best to keep several of these fishes in the same tank, as they are very social in their habits, and will not live long solitary.

The minnow (*Cyprinus*, or *Phoxinus lævis*) is another of our commonest little fresh-water fishes,

Fig. 21.

Minnow (*Phoxinus lævis*).

and remarkable for assuming prismatic colours during the spawning season, after the manner of the male sticklebacks. Like the gudgeon, it is gregarious, and usually found in shoals; but, unlike it, it is fonder of the surface, where it can bask in the sunshine, than of the bottom of the water. It lives as well in confinement as the sticklebacks do, and is quite as vivacious and attractive. Indeed, many people prefer it to the stickleback, on account of its more peaceful habits.

The pope (*Perca fluviatilis minor*) is a beautiful fish, a well-marked variety of the common perch.

Although not so easy to preserve in the aquarium as
the foregoing, a little extra attention will be sufficient
to keep it in good health for a long period of time.

Fig. 22.

Common Perch (*Perca fluviatilis*).

It grows from 4 inches to 6 inches in length, and
forms a very conspicuous object, moving to and fro
over the bottom of the tank in search of caddis-
worms, blood-worms, crustacea, &c. It loves to swim
boldly about with its dorsal fin erected, and then
looks as if engaged upon some crusade.

The common carp and Prussian carp (*Cyprinus carpio* and *Cyprinus gibelio*) are both fishes which may be easily kept in our larger fresh-water aquaria.

Fig. 23.

Common Carp (*Cyprinus carpio*).

The former grows to a great size, but might easily be removed when it is too big for its tank. Both these fishes live to a great age, and are very tenacious of life, the latter especially so. When of the length of three or four inches the common carp is a very useful and even an attractive fish. It is a good scavenger, and

feeds equally on vegetable and animal food. When
young the bronze tinge gives them a very pleasing
appearance. During the winter their habits are very
sluggish, even when the aquarium is kept within
doors; and then they eat next to nothing, but pass
the time in a semi-hybernation. The Prussian carp
is even prettier than the above-mentioned species, possessing a brighter colour and a more graceful shape.
It is very fond of warm water, and breeds in enormous
numbers in the fresh-water "lodges" or reservoirs into
which the steam-water flows from manufactories in
the north of England. No more hardy or useful fish
could be placed in the tank than this. The bleak
(*Cyprinus alburnus*) is another interesting and useful

Fig. 24.

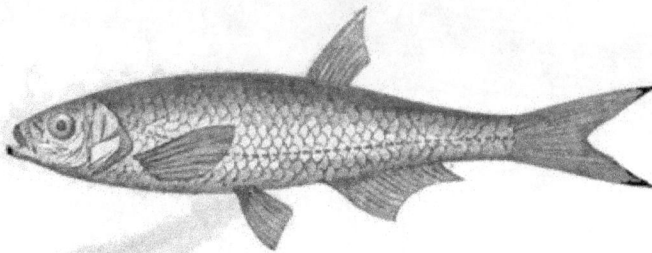

The Bleak (*Cyprinus alburnus*).

fish, one that will give little trouble, and which,
in the larger tanks, will prove of great service in
clearing away decomposing organic matter of all
sorts. Not long ago the scales of this fish were used
and ground up for the manufacture of artificial pearls.
The shape of these fishes is very graceful. They

F

usually move in shoals, hence it is necessary to keep
several of them together.

The roach (*Cyprinus rutilus*, or *Leuciscus rutilus*)
is an attractive object in the aquarium, although

Fig. 25.

The Roach (*Cyprinus rutilus*, or *Leuciscus rutilus*) and Dace
(*L. vulgaris*).

more difficult to keep than many of those we have
mentioned. When young, and free from the black
cancerous growth on the scales which disfigure so
many of the older and larger fish, it is so active and

healthy that we cannot wonder at the common saying,
"As sound as a roach." Its bright red fins assume
their most gorgeous tints about breeding time. A
nearly allied species, common in Norfolk, is the rudd,
so called on account of its still brighter tints of red.
When young this species may be domesticated, and
kept healthily alive for some time, especially if the
surface of the water be covered with aquatic vegetation, for it is very fond of hiding under the cool
shelter of such a layer, and darting out thence suddenly and vigorously on its prey.

In the larger tanks the perch may be kept with
comparative ease, although it is a fish very capricious as to diet. It is always best to begin with
small specimens, or those that are about a quarter
grown. A perch of two or three ounces weight is
quite big enough for ordinary size tanks, especially if
other animals are kept with it. In the larger ones

Fig. 26.

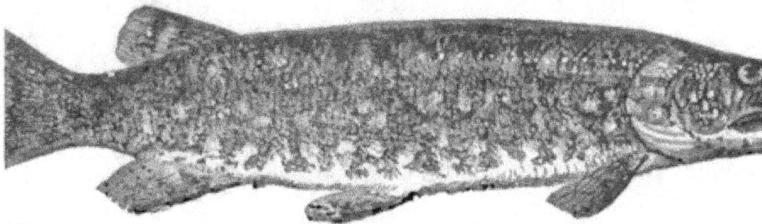

The Pike (*Esox lucius*).

small pike (*Esox lucius*) may be preserved, but Dr.
Lankester calculated that a pike which weighs a
pound will require a tank containing 32 gallons of

F 2

water, besides a forest of aquatic plants to decarbonise
and oxygenate it. This estimation is perhaps too high,
but there can be no question of the large quantity of
water required, especially in tanks where there is not a
constant change by circulation. The pike may be fed
with minnows or small roach, which will swim about
until captured. This generally takes place in the
early morning, for pike love to bask in the mid-day
light, and are then so oblivious of external things that
one of the modes of capturing them is by passing a
running noose of copper wire over their heads whilst
they are thus meditating. Much has been said respecting the acclimatisation of the *Silurus* in this
country, and the general establishment of aquaria
might tend to settle the question of its practicability.
This fish is tolerably common in Germany, Norway,
and Sweden, where it grows to a large size. Many ich-
thyologists have believed that it could be introduced
with advantage into many of our rivers and lakes,
where it might be cultivated as an additional staple
of human food. Its flesh is said to be of excellent
flavour when properly cooked. As the long barbules
near the mouth indicate, it is a bottom feeder. It is
stated to attain the weight of fifty-six pounds in
four years, in waters where food is abundant. The
longest of the barbules seem to be put to the same use
as the dorsal spines are with its marine representa-
tive the angler, or fishing frog (*Lophius piscatorius*).
Yarrell, who wrongly gives the silurus a place among

British fishes, states that it is sluggish in its habits and a slow swimmer, and that it takes its prey by lying in wait for it like the fishing frog. From the extensive geographical distribution of this fish there can be little doubt it might be acclimatised in English waters. In 1865, fourteen live specimens were introduced to the Thames, and subsequently Mr. W. A. Lloyd introduced many more, but we have not heard of their fate.

Another fish, not uncommon in North American lakes and rivers, might be introduced into our larger fresh-water aquaria, and be the means of enlightening the general public as to the nature of by far the larger number of fishes which lived during the Primary epoch, and whose fossil remains occur so abundantly in the Devonian and carboniferous rocks. During those epochs this class of fishes reached its maximum, and ever since has been slowly dwindling away. Only a few species, forming about three per cent. of all the fish fauna of the world, remain to represent a once dominant and cosmopolitan race. We allude to the *Ganoid* fishes, so-called from their bony scales being covered externally with a glossy enamel. The sturgeon belongs to one division of this interesting group, and, notwithstanding its migratory habits, the latter seems to have done well in the Brighton and other aquaria. We are not aware whether the North American gar, or bony pike (*Lepidosteus osseus*), has yet been introduced into England. It is a gracefully

shaped fish, and we have seen specimens nearly
3 feet in length, whose whitish enamelled plates re-
minded us strongly of what many of the ancient
primary fishes must have been. The habits of this
suggestive fresh-water fish very much resemble those
of its English namesake, hence, no doubt, the reason
why it was called " gar-pike " by the colonists. We
hardly need point out the advantages which large
aquaria possess in making us acquainted with the life-
history of desirable fishes, before we attempt their
acclimatisation for ulterior purposes. Hitherto we
have done this quite at a venture.

CHAPTER VI.

THE AQUATIC GARDEN AND ITS PLANTS.

UNTIL within a year or two ago the circulatory system of aerating large aquaria was confined to salt water. After Mr. Lloyd applied it thus, at Hamburg, in 1868, Mr. Saville-Kent applied it to the fresh-water tanks in the Manchester Aquarium with some success. The chief means of aerating the water was either by means of fountains, or through the agency of aquatic plants. Of course it will be impossible to amateur keepers of aquaria to aerate their small reservoirs by mechanically circulating the water. Such a process would be troublesome, and would hardly repay the large expense and outlay of labour by the small returns that would be afforded. Moreover, one reason for keeping aquaria in rooms is on account of their pretty and attractive appearance, and this would not be so effective without the greenery of aquatic or semi-aquatic vegetation which they usually possess. It is true Mr. Kent found that fishes suffering from the attacks of a white fungoid growth quickly recovered when placed in tanks aerated by circulation. Mr. Kent is of opinion that the fungoid growth is due to the *lime* in the water.

In this opinion we cannot agree. First, we are not aware that lime is at all an adjunct to the growth of microscopic fungi, and the Manchester water is singularly free from it as compared with that of most towns. We believe the fungoid growth to be due to a surplus of *nitrogenous* matter in solution, caused, perhaps, by the decay of animal substances to which tanks are liable if not properly attended to. Nitrogen is a necessity and stimulant to the growth of fungi of all kinds, and it is evident that the removal of affected fishes to clear running water where nitrogen is absent would counteract the disease. It has long been known to keepers of fresh-water aquaria that the best way to cure fish affected by the white fungus was to remove them to a tub where the water was constantly dripping from the tap. We have referred to this matter at length because it is one that cannot fail to interest aquarium owners.

A judicious selection and grouping of aquatic vegetation adds greatly to the beauty of a fresh-water aquarium; and we have hinted how plants requiring deep soil, as well as those requiring little or none, may be accommodated in the same tank, by planting the former in ordinary flower-pots. These may be so arranged along one or other of the sides as to be only just covered with water, and then it will be possible to grow such semi-aquatic plants as the lovely buck-bean (*Menyanthes trifoliata*), sun-dew (*Drosera rotundifolia* and *anglica*), *Polygonum amphibium*, the

flowering rush (*Butomus umbellatus*), the water mint
(*Mentha aquatica*), mare's tail (*Hippuris vulgaris*),

Fig. 27.

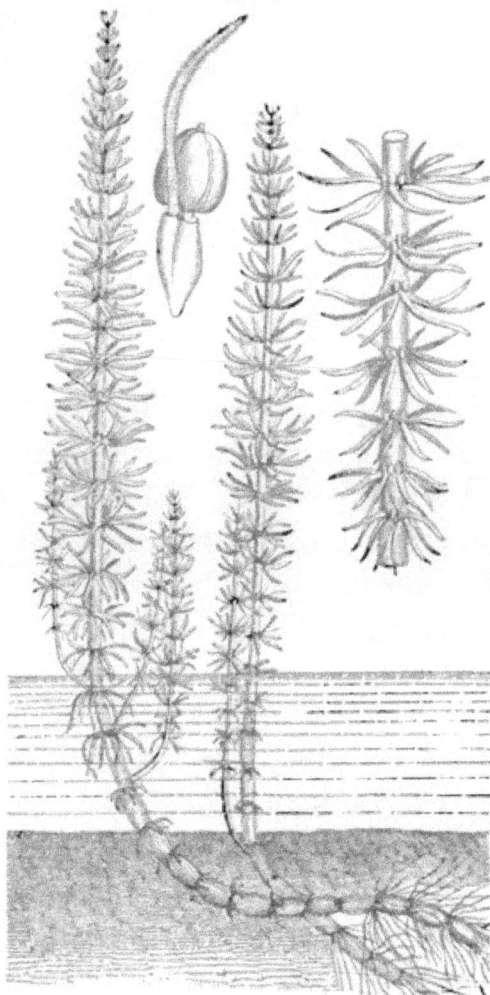

Mare's Tail (*Hippuris vulgaris*).

and others. Our native vegetation offers considerable choice, and for the beauty of the foliage and

flowers of some even of the commonest species, it would be difficult to find an equal number of foreign plants that would excel them in this respect. It should be remembered, however, that the *main* thing in the selection of aquatic plants is to have such species as will most effectively oxygenate the water. And it is always best to have the vegetable element slightly in excess of the animal, inasmuch as it can be more effectually managed, either by elimination, or by moderating the light which stimulates its growth. The common American weed (*Anacharis alsinastum*) has never yet been known to flower or seed in this country. It is propagated solely by budding, and we have already said that fish, and especially goldfish and carp, are very fond of nibbling the young and tender shoots. A little of this plant may therefore always be advantageously introduced into a tank of moderate capacity. It must be seen, however, that its tendency to rank growth does not interfere with other and more ornamental plants.

Another species which has long been utilised in this country by druggists, to oxygenate the water in which medical leeches are kept, is the *Vallisneria spiralis*. It is a native of the south of Europe, but has been pretty extensively distributed in England. This plant is undoubtedly one of the best and most copious yielders of oxygen of all common fresh-water species, and therefore is of great service in small tanks. Its grass-like leaves show the circulation of

the protoplasmic granules under the microscope, like
the blood corpuscles in the web of a frog's foot. This
plant roots freely in a little sandy earth or mud. The

Fig. 28.

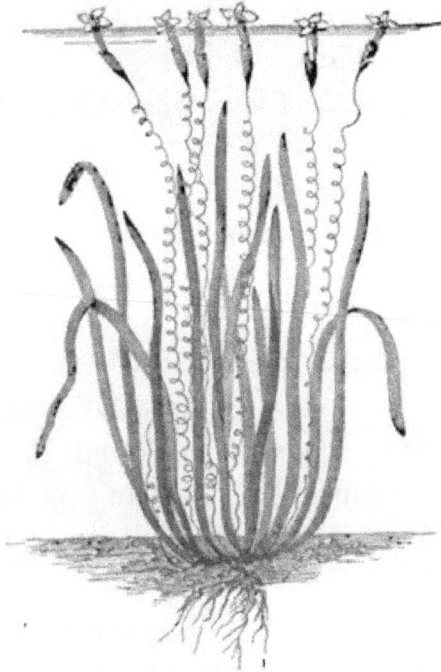

Vallisneria spiralis.

male and female flowers are borne separately, the
latter having the long spiral stalks which have given
to the plant its specific name. These float on the
surface of the water; whilst the male flowers are
borne on short stalks at the base of the plant. They
are detached thence just before opening, and rise to
the surface to fertilise the female flowers with their
pollen.

Few flowers exceed in beauty the yellow and white
water lilies of our English streams and lakes (*Nuphar
lutea* and *Nympha alba*). The space they take up,
however, is so great that they can only be introduced
into very large tanks, or into the basins of garden
fountains. Their broad, cool leaves form an admirable
retreat and screen for fishes, especially in the heat of
a summer's noonday. The perfume of the flowers of
the former plant has obtained for them the name of
"brandy-bottles" in some parts of England. The
flowers only rise above the surface and open in the
full blaze of day—at night the petals close and the
flower head is withdrawn into the water. The water
plantain (*Alisma plantago*) is a pretty aquatic plant,
whose panicles of pale liliac flowers rise above the
water very prominently. Its name is derived from
the resemblance of its leaves to those of the common
plantain. The flowers secrete honey from twelve
glands they possess, for the purpose of attracting
insects to bring about the cross-fertilisation which
is so important an element in the propagation of
many flowering plants. The arrow-head (*Sagittaria
sagittifolia*) belongs to the same order as the last-
mentioned species, and is another true aquatic plant
which may be included in our list of those intended
mainly for ornamental purposes. Its common name
is derived from the conventional arrow-head shape of
the bright green leaves. The flowers are very pretty
whilst they last, being white with pinkish spots at the
base of the petals. The upper ones bear stamens

only, and the lower pistils. In this order of plants also we should not forget to mention the flowering

Fig. 29.

Water Plantain (*Alisma plantago*).

rush (*Butomus umbellatus*), one of the prettiest of all our native species. It requires to be planted in soil

placed in a flower-pot, and hidden away in rockwork,
as already described.   The bright rose-coloured flowers

Fig. 30.

Arrow-head (*Sagittaria sagittifolia*).

then rise from amid the three-cornered, sword-shaped
leaves in large flat umbels.   The root of this lovely

plant is believed in Russia to be a cure for hydro-
phobia, but evidently without reason.

Fig. 31.

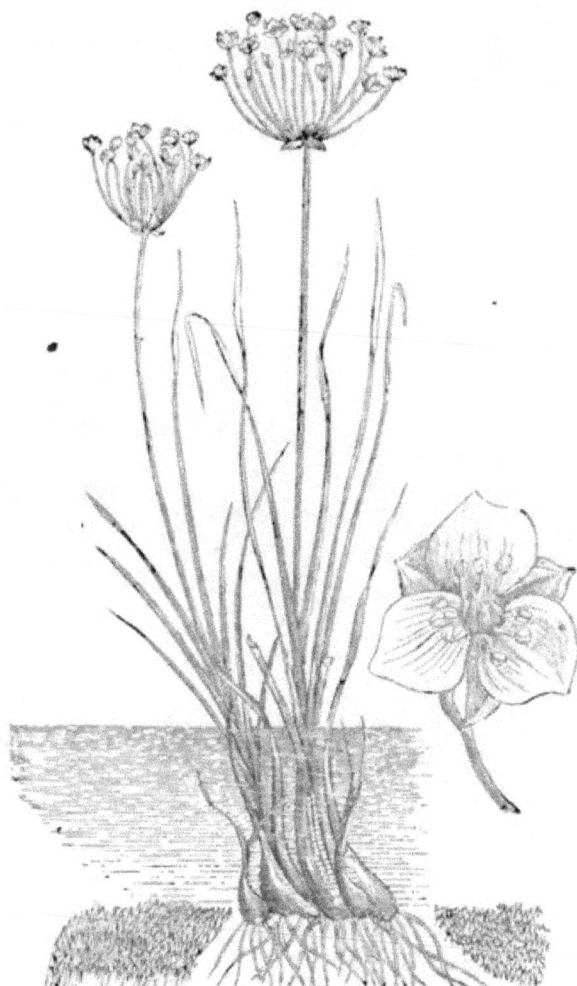

Flowering Rush (*Butomus umbellatus*).

Among other native or ornamental plants which
may easily be adapted to the aquarium are the marsh

forget-me-not, water mint, brook lime, water violet, polygonum, and buck-bean. The first (*Myosotis palustris*), also goes by the name of "scorpion grass."

This is the veritable forget-me-not of the legend, and forms an exquisite adjunct to an aquarium, especially as a surface bordering. The flowers are coiled up in the crozier shape peculiar to the unopened fronds of ferns,

Fig. 32.

Water Mint (*Mentha aquatica*).

before they expand. The water mint (*Mentha aquatica*) gives out a grateful perfume, especially in the

evening, whilst its clusters of pale liliac flowers are
unobtrusively pretty. The brook lime (*Veronica becca-
bunga*) is noticeable for its bright green leaves and

Fig. 33.

Brook Lime (*Veronica becca-bunga*).

deep blue flowers. A nearly allied species is the
water speedwell (*Veronica anagallis*), whose flowers
are paler, and rise above the water in loose panicles.
The brook lime has to be looked after, otherwise it de-
velops too rapid and rank a growth. The water violet
(*Hottonia palustris*) is another of our prettiest English
plants, as anyone will acknowledge who has seen the

G

ditches full of it in Norfolk and Suffolk. Its submerged leaves are feathery, like those of the water

Fig. 34.

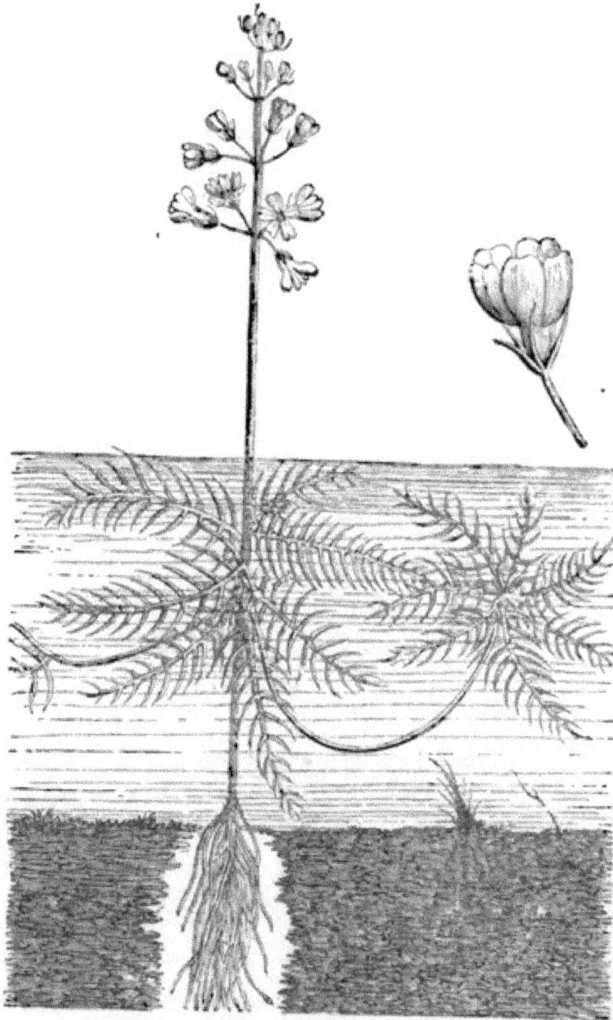

Water Violet (*Hottonia palustris*).

ranunculus, only larger and prettier; whilst its upright hollow stem is surrounded with whorls of deli-

cate five-petaled, liliac-coloured flowers. *Polygonum amphibium* is an aquatic species which may be easily transferred to a tank, and which readily flourishes there. Its bright, light-red flower spikes raise themselves just above the semi-floating leaves, and look very pretty indeed. But of all our British plants affecting moist habitats, surely none is so lovely as the buck-bean (*Menyanthes trifoliata*). It will not

Fig. 35.

Buck-bean (*Menyanthes trifoliata*).

flourish in the water, but may be so potted as to grow on the margin, at the top of rockwork which may be arranged at one end of the tank. In this position its bright deep green trifoliate leaves, and pale pink and white, feathery, lace-like flowers, cannot fail to obtain sincere admiration. In Lancashire this plant is much esteemed and gathered by the common people, the intense bitter of the leaves being esteemed a good stomachic. In Sweden and Norway the leaves are used in brewing beer, instead of hops. The plant has the floral peculiarity of bearing two sets of flowers, and is therefore *dimorphous*.

We now turn to those common plants which are more distinctly aquatic than some of those we have just noticed. Of these by far the most widely distributed is the water crowfoot (*Ranunculus aquatilis*). It bears two sets of leaves, those intended for floating on the surface, which are broad and slightly lobed ; and others which are always immersed. The latter are the most numerous, and are thread-like and feathery, often growing so fast as to form dense tangles. The flowers are pure white, with numerous yellow stamens. They rise singly above the water, and may be seen covering our ponds with a perfect carpet of flowers. The water ranunculus is a famous hiding place for aquatic objects. Water spiders build their queer, "diving-bell" like nests amid its leaves ; insect larvæ hide in the depths of its thickets ; rotifers attach their frail cases to its tiny threads ; fishes of all kinds lie in wait or hide from enemies, or cool themselves beneath its dense and almost impenetrable foliage. No aquatic plant is more useful for the aquarium, or trained with greater ease ; but, like the *Anacharis*, it has

to be prevented from usurping the entire tank. Even
more beautiful than this is the frog-bit, whose kidney-
shaped, bronze-coloured leaves are most exquisitely

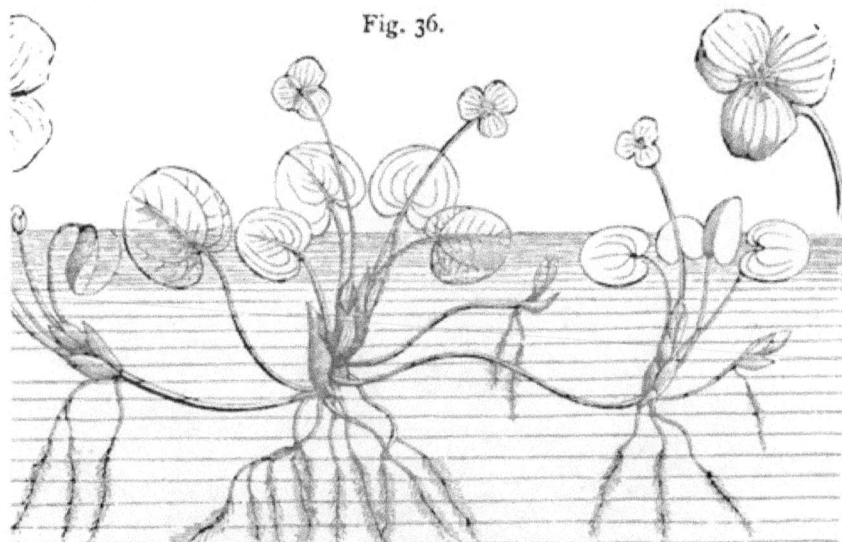

Fig. 36.

Water Frog-bit (*Hydrocharis morsus-ranæ*).

meshed with veins. The plant is very tenacious of
life, and may readily be accustomed to the aquarium.
The flowers possess three petals, and are white and
fragile looking. The water soldier (*Stratiotes aloides*)
is a very abundant plant in Norfolk and Suffolk, as
anglers thereabouts have long since discovered. It
has to be planted in deepish soil at the *bottom* of the
tank. Its leaves are thick and strong, and fringed
with a margin of recurved spines like those of the
aloe, whence its other name of the "water aloe."
The flowers are very pretty, and have three petals

like those of the frog-bit. These are borne to the top
of the water when ready, in order to be properly fer-
tilised. As the plant is of slow growth, and tolerably

Fig. 37.

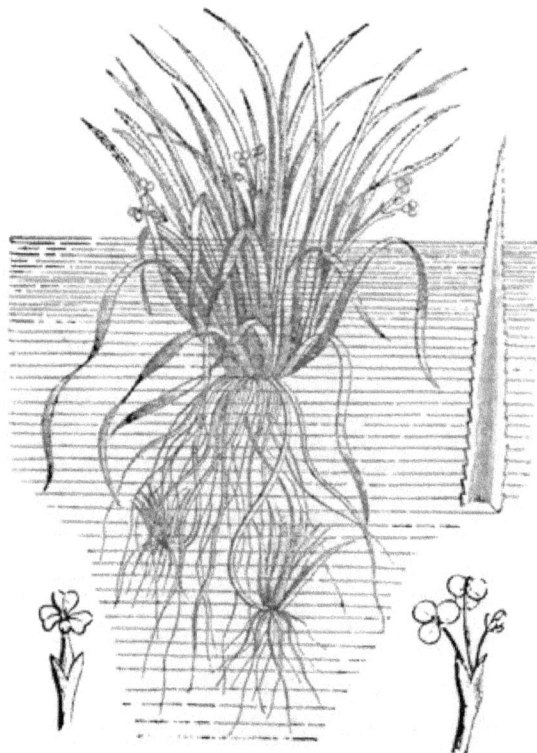

Water Soldier (*Stratiotes aloides*).

hardy, it may advantageously be used for aquarium
purposes, although it is not of much service for
aerating the water. Another and a rarer British plant,
but one which is not uncommon in the sluggish rivers
of the east of England, is the bladder-wort (*Utricu-*

laria vulgaris). The submerged leaves are thread-like, as is so usual with those of aquatic plants, owing to the greater ease with which they can thus be brought

Fig. 38.

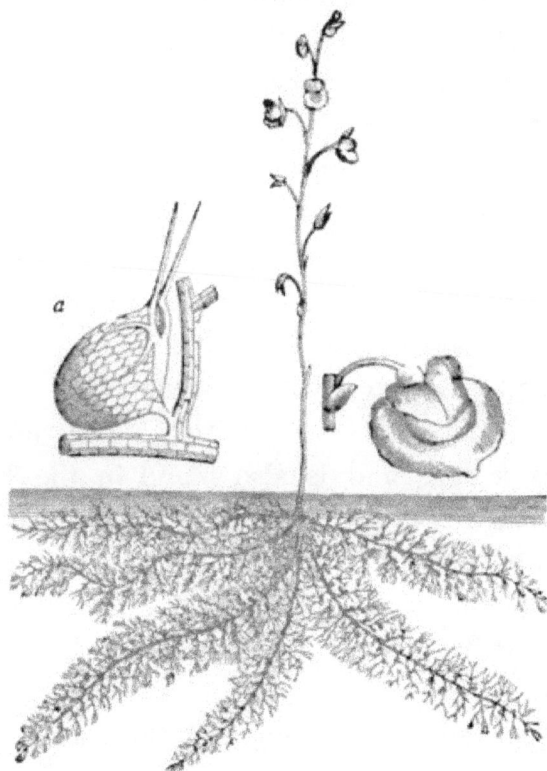

Bladder-wort (*Utricularia vulgaris*).

into contact with fresh supplies of water. The purple flower-stalks rise above the surface, and bear bloated looking *yellow* flowers, having an upper and a lower lip, the former being further adorned with purplish veins. The common as well as the scientific name of this pretty and interesting plant is derived from the presence of certain little bladders attached to the thread-like leaves (*a*). These fill with air, and buoy up the flower-stalks above the water. After fertilisation has been effected they fill with water, and thus the entire plant sinks so that the seeds can be ripened at the bottom. The bladder-wort is therefore one of the most interesting of all our aquatic plants, whilst its yellow flowers appear very prominent amidst the whitish or pinkish tints which our British aquatic flowers are usually adorned with.

Villarsia nymphæoides is a much rarer plant than many of the above mentioned, but it is an exquisite object, and may no doubt be obtained living from London dealers. It is a British plant, inhabiting ponds as a rule, and is also found in the Thames. The leaves are round, and float on the top of the water; whilst the largish yellow flowers are borne on single stalks. *Trapa natans*, although not a British plant, is easily procurable; as it is found abundantly in many European streams and rivers. Its flowers are whitish, tinged with red, and the leaves grow in very elegant semi-floating clusters. We have already referred to the ubiquitous pond weeds (*Lemna*) as forming a fresh, green covering for small aquaria. It is not wise to cultivate this plant too much, as it gives a rather poverty-stricken look to the whole tank. Among other plants which are useful for oxygenating purposes we may mention the water star-wort (*Callitriche verna*), the horn-wort (*Cerato-phyllum demersum*), and the pond weeds (*Potamo-geton*), of which we have several abundant species,

Fig. 39.

Villarsia nymphæoides.

easily procurable. The water milfoil (*Myriophyllum spicatum*) is a very common and graceful aquarium plant. Its greenish flower spike is often raised un-obtrusively above the water; whilst its submerged thread-like leaves are arranged in whorls of four in number. This is the commonest of our British species of milfoil. The water star-wort (*Callitriche verna*) has long been extensively used in fresh-water

Fig. 40.

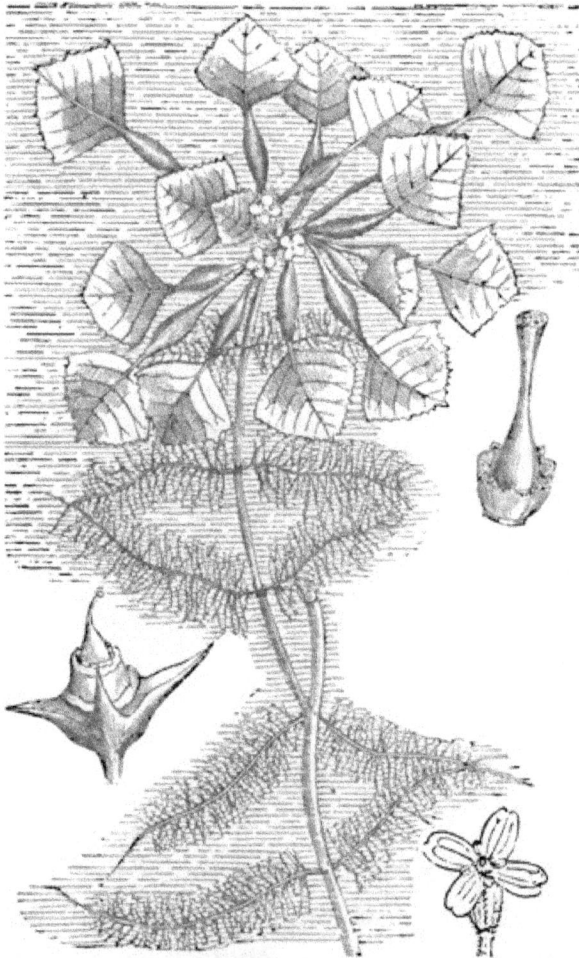

Trapa natans.

aquaria, not only on account of its aerating powers, but also because of its pretty green leaves, and the fondness which many aquatic animals manifest for employing them in several ways. The leaves are small and arranged in regular star-shaped or rosette-

Fig. 41.

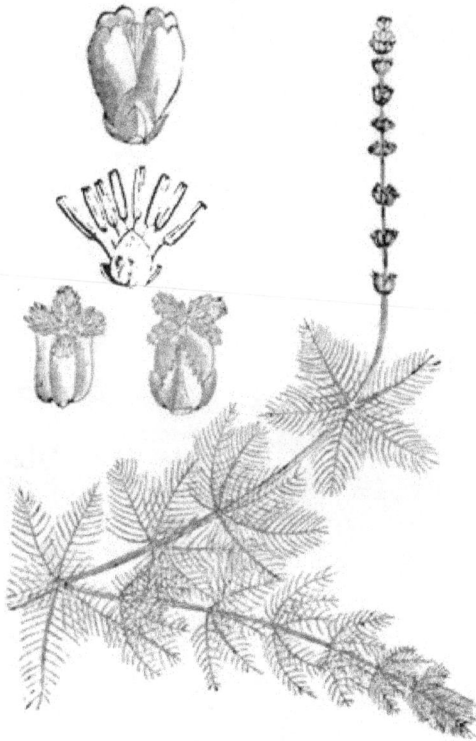

Water Milfoil (*Myriophyllum spicatum*).

like whorls; whilst the very simple flowers are green-ish, and easily passed over. The upper leaves may usually be seen floating on the water, arranged in the star-like form which has obtained for the plant its common name. Of all the pond weeds perhaps *Pota-mogeton natans* is the prettiest and most serviceable for aquaria.   The flower spikes are densely set with small reddish flowers, and are slightly raised above the surface of the water.   The leaves are brownish-

Fig. 42.

Water Star-wort (*Callitriche verna*) ; *a, b,* flowers.

green, and very thin and prettily veined.   These leaves are usually submerged, but few of them persistently floating on the top.

Besides these aquatic plants, useful for aerating

the water of the aquarium, many others might be
mentioned which could so be planted as to lend an
additional charm to the exhibition of living animal

Fig. 43.

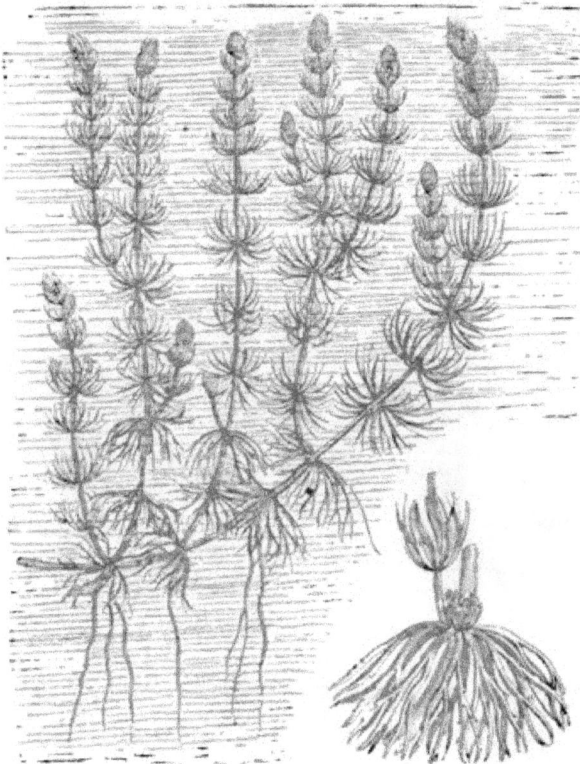

Horn-wort (*Ceratophyllum demersum*).

and vegetable forms. Among these we recommend
the sweet flag (*Acorus calamus*) and the bur reed
(*Sparganium ramosum*) as being not only the most
easily procurable, but as forming the most graceful
adjuncts. They may easily be planted within the

tank, in deepish soil or in flower-pots, and then,
mixed with the water violet or water plantain, they
form a varied and agreeable bordering.   The singular

Fig. 44.

Sweet Flag (*Acorus calamus*).

looking balls of flowers on the stems of the latter
plant cannot fail to draw attention, whilst the elegant
linear leaves of the sweet flag are so fragrant that
they were once used for strewing on church floors on
that account.

We have endeavoured to make such suggestions
respecting our native aquatic plants as will not only
render them easily identifiable, but enable the amateur
to utilise some of them for the purpose of aerating

his aquarium. Many of them we have either grown
ourselves, or seen growing under conditions similar to

Fig. 45.

Bur Reed (*Sparganium ramosum*).

those described. At the same time we have ventured
to hint that such selections might be made as would

convert the aquarium, if the tank be large enough, into an aquatic garden as well. No British plants have such a fragile hot-house look as our aquatic species, and they would therefore suit the life of indoors admirably, and be the means of contributing another element to the many surroundings which already make our English homes the happiest on the earth !

CHAPTER VII.

MOLLUSCA, INSECTS, ETC., OF THE FRESH-WATER AQUARIUM.

THE solitary naturalist in his search after the manifold living forms of life soon feels, as William Cullen Bryant says in "Thanatopsis," that

> " To him who in the love of nature holds
> Communion with her visible forms, she speaks
> A various language."

It is a language which gives forth no uncertain sound ; and, although the mystery of earthly life starts forth even more vividly when the student discovers the hourly carnage by which it can alone be sustained, this does not detract from an unshaken confidence in the wisdom and even love of the Almighty Power that superintends it ! Mere earthly life is not the highest thing in the universe. The carelessness with which myriads are crushed, and even their types are lost, proclaims it to none more clearly than to the naturalist. We see these things but as in a glass darkly, yet we obtain a glimpse of the important fact that the life-scheme of our globe, past and present, is one and indivisible, and that the individual members of it which perish and give place

H

to others, have no more right to complain than the
blood-corpuscles of our body, when they are spent in
energy and replaced by those newly formed!

The aquarium keeper soon finds that it is necessary to be constantly replenishing his stock.   Not
only has he first to get something like a balance of
animal and vegetable life, he has also to see that
the associated animals do not breed too fast or too
slowly.   If the former, then he introduces one or two
species which keep them down by preying upon
them ; if the latter, he adds additional specimens.
If all the animals of his aquaria be carnivorous and
none herbivorous, his tank will soon be converted into
a regular field of battle, and war will be the order of
the day, until the combatants are reduced to the
fabled condition of the Kilkenny cats.   He has,
therefore to copy nature in this respect, and mind
that his aquatic pets are taken from the carnivorous
and herbivorous classes alike, and that they are
placed in such a relation to each other that the
marvellous fecundation of the herbivorous group re-
places the ravages made upon them by the carnivorous.   It is evident that if the herbivorous kinds
are in the ascendant, harm will soon issue to the
balance of life by the oxygen-yielding plants being
devoured.   Hence the importance of having both
carnivorous and herbivorous creatures in the same
tank if possible.

We have dwelt at some length upon the commoner

fresh-water fishes and amphibians with which the ordinary aquarium may be stocked ; and now briefly refer to other species not belonging to these orders, which are equally common in all our ponds and tarns, and equally interesting and animated when transferred to the aquarium. First of all we may notice those commoner species of water snails, whose hardiness and voracity prove of great service in keeping down the impalpable green algæ, which will develop even in the best managed aquarium. Of

these none are more abundant than the *Limnaceæ*, and of the eight British species included in the order, perhaps the best is *Limnea stagnalis*. It will crawl over the inner surface of the glass, and keep it as clean as if well dusted. Occasionally it may be seen floating, and then before it descends to the bottom of the tank, it utters a perceptible sound, caused by disengaging the air from its pouch, which had kept it buoyant. This species is the handsomest we

Fig. 46.

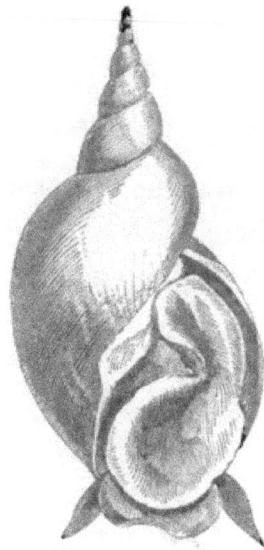

Lymnea stagnalis.

have, and the young shells are especially graceful and slender. *Limnea auricularia* is a much smaller shell, with a larger body-whorl, the outer lip of which is reflected. Although not so common as *L. stag-*

H 2

nalis and *L. pereger*, it is far from rare. Its habits
are very similar to *L. stagnalis*, except that it is
fonder of confervoid vegetation ; and as this always
tends to become a pest, *auricularia* is therefore a
valuable addition to a fresh-water tank. *Stagnalis*
has an undue preference for certain of the higher forms

Fig. 47.   Fig. 48.

Limnea auricularia.

Limnea pereger.

of aquatic vegetation, notably *Vallisneria ;* and if
there be any of this plant present it will surely feed
on it in preference to any other. When *Vallisneria*
is too rank in its growth these molluscs may therefore be employed to temporarily keep it down.

Limnea pereger (Fig. 48) is the most abundant of all
our native species, and, as is commonly the case with
species that are individually numerous, it has given
rise to at least fourteen well-marked varieties. It is
more active in its movements than the species already
noticed, and likes occasionally to creep out of the
water up the stems of aquatic plants for a temporary
breath of fresh air. It is exceedingly prolific, Dr.
Gwyn Jeffreys stating that it lays about thirteen

hundred eggs, in numerous clusters. It is further
distinguished from its specific brethren in not being
confined to a purely vegetable diet, but occasionally,
sometimes frequently, indulging in a meal off a dead
fish, or even brother or sister.

· The genus *Planorbis* is not geologically older than
Limnea, both having an antiquity which dates back
from the early tertiary period. We have eleven

Fig. 49.

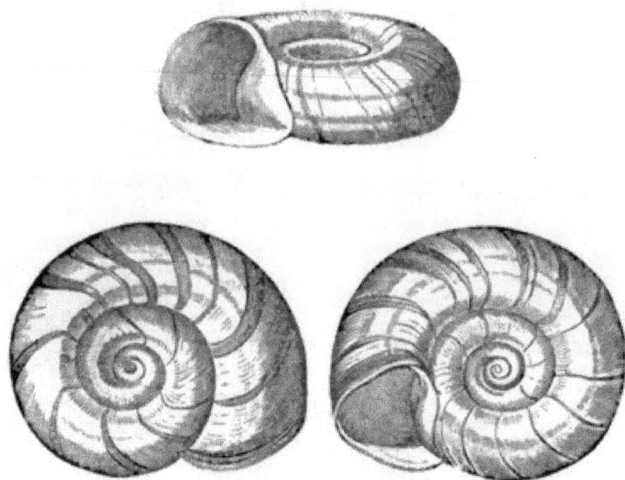

Planorbis corneus.

British species, most of which are to be met with in
every river, canal, pond, or tarn. They vary considerably in size, some being scarcely larger than a pin's
head ; whilst *Planorbis corneus* often measures three-
quarters of an inch across. All of them are herbivorous, and have equally the same habit, when crawling,
of lugging their shells behind their half-extended

bodies. *Paludina* is a genus geologically older than
either of the foregoing; for the remains of certain
species of this shell completely make up the bulk of
the fresh-water marbles of the Wealden and Purbeck
beds. Like most fresh-water genera, the forms have
not greatly varied, owing, perhaps, to the similarity
of fresh-water conditions in the most remote periods
to those of the present time. We have two species
extant in our ponds and rivers, of which one, *Paludina
vivipara*, has always been a great favourite with

Fig. 50.

Planorbis
marginatus.

Fig 51.

Paludina vivipara.

aquarium keepers. As its specific name implies, it
generally brings forth its young alive; that is to say,
it keeps the eggs within its body until they are hatched
there. This, however, is not always the case. The
males are said to be usually smaller than the females.
Both are tolerably active, and look very pretty with
their colour bands running up the whorls of the shell.
Both in this genus and another nearly related to it,
which is even commoner in our ponds and streams,

the *Bythinia*, the mouth of the shell is protected by an *operculum*, or door. *Bythinia* is much smaller, and of a more social character, so that we usually find it in great numbers. It also makes a good aquarium object, and lays its eggs in three long rows, usually on stones if there be any, or if not on the stems and leaves of plants. Bivalve shells, such as the little *Sphærium* and *Pisidium*, may also be safely introduced. Their habits are active, espe-

cially those of the ubiquitous *Sphærium lacustre*, which crawls up and down the tank as if it had only one shell instead of two, and occasionally indulges itself in a waltz, re-

Fig. 52.

Sphærium corneum.

volving at the rate of fifteen or twenty circles a minute! The tiny pea-shells (*Pisidium*) are also very abundant, and are extremely useful in an aquarium from their scavenging habit of devouring any dead and decomposing animals, of which diet they are very fond. If the tank be large enough, and the soil at the bottom sufficiently deep, a specimen of the great swan mussel (*Anodonta cygnea*) might be transferred. Its half-opened shell, showing the tufted syphons, makes it a very pretty and interesting object; and the student can witness the currents created by the ciliated tubes, by which fresh air and food are taken in, and effete water and matter carried out. A species of *Unio* might be used instead in smaller tanks, as this is a smaller

bivalve shell.   The numerous eggs of both these common bivalves will furnish the fish with abundant food.

The aquatic insects that every pond usually swarms
with, are not less attractive and interesting than the
fishes and mollusca.   Indeed, some of them surpass
the latter in interest, on account of the life-stages
through which they pass ; such as the larvæ of the
dragon-flies, caddis-worms, water beetles, &c.   What
could be more astonishing than the fact that the early
life of many aerial and winged insects is passed in
water, under conditions which are as contrasted as
possible with those which affect them in the adult
condition ? Even in their individual life-histories these
creatures furnish a sufficient answer to those who
demand "missing links!"   And it is not a little
suggestive that the insect orders which appeared first,
during the carboniferous epoch, were those whose
members now pass through *less* differentiated larval
stages than those which were introduced later on, such
as the butterflies (*Lepidoptera*) and *Neuroptera.* Moreover, the *Orthoptera* (a fossil species of which is the
first to appear of all known kinds of insects) is an
order which perhaps even yet furnishes a larger
number of species whose lives are passed under
aquatic as well as aerial conditions, than any others.
Even the caddis-worms, which are the larvæ of insects in some respects nearly related to the Lepidoptera, have an enormous geological antiquity.   We
have seen limestone beds 5 and 6 feet thick, com-

posed of their shelly tubes alone, in central France. These are of miocene age.

The young of the dragon-flies do not assume a quiescent attitude when they pass into the pupa stage. On the contrary, they are more active and voracious than ever. Any old pond or tarn will yield these insects, and they may afterwards be watched going through those evolutions which eventually end in the insect leaving the water and climbing up some water plant so that it can shake off its old clothes, and enter upon a winged existence. Perhaps, when living in the water, the larvæ of the dragon-fly are never more

Fig. 53.

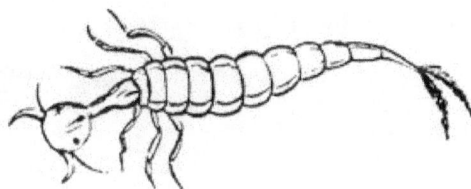

Larvæ of *Dyticus.*

actively engaged than when chasing the water fleas (*Daphnia*). The larva of the *Dyticus* is still more ravenous, and is so courageous and fierce, that it has earned the popular name of the "water devil." It will attack sticklebacks, minnows, the larvæ of dragon-flies and water scorpions; whilst the poor tadpoles are sacrificed by scores to its hungry maw. Nay, so fearfully are they afflicted with the sensation of hunger that they will fall on each other, if nothing else be present.   They frequently cast their skins, which may be mounted as microscopical objects, showing the spiracles, or breathing mouths, peculiarly adapted to water. So fierce are these common objects that they will readily seize and hold on to a stick with whose end they have been taunted. The *Dyticus marginalis* varies so considerably in the appearance of the sexes that they are not unfrequently mistaken for two different species of beetles. Their hind-legs are peculiarly adapted for swimming, owing to their

Fig. 54.

Water Scorpion attacked by Larva of *Dyticus*.

Fig. 55.

Male of *Dyticus marginalis*.

Fig. 56.

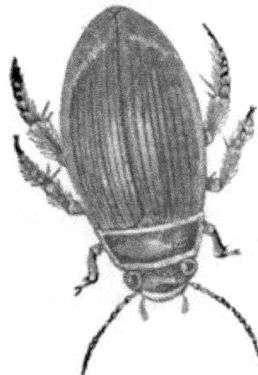

Female of *Dyticus marginalis*.

being flattened out into oar-like expansions. The fully-developed insect bears out the character it obtained in its larval stages as the "water devil" on account of its voracity and destructiveness. Indeed, it is a very difficult matter to keep the larvæ, unless we separate them. The fully-developed *Dyticus* leaves the water during the night, in the summer months, so that if the top of the aquarium be not closed, its owner may be somewhat astonished at seeing some of its aquatic inhabitants leaving it to dash themselves against the gas-globes!

Fig. 57.

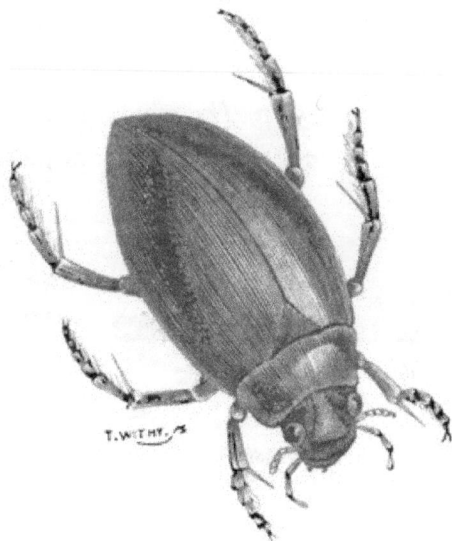

Great Aquatic Beetle (*Hydrophilus piceus*).

Hydrophilus or *Hydrous piceus*, or great aquatic beetle, is the largest of our British species. It is wondrously contrastable with the *Dyticus* in its habits, for it is extremely inoffensive, and therefore well suited to a fresh-water aquarium. Indeed, the full-grown insect not unfrequently falls a victim to the

savageness of the larva of *Dyticus*. It is interesting
to watch the *Hydrous* lay its eggs, in a kind of silken
cocoon, spun by the mother.   In this cocoon the eggs
float about until they are hatched.   Curiously enough
this beetle swims by *alternate* movements of its legs.

The smooth surface of most ponds may often be
seen streaked by the mazy paths of the whirligig
beetles (*Gyrinidæ*).   Other and not uncommon water
beetles are the *Colymbetes*, related to the *Dyticus*.

The *Hemiptera*, or water bugs, are represented in
most ponds by the water scorpion (*Nepa*), and water
boatmen.   The former (which is engraved in Fig. 54
as being attacked by a *Dyticus* larva) is itself so
voracious as to have obtained the popular name it

Fig. 58.

Fig. 60.

Fig. 59.

Fig. 61.

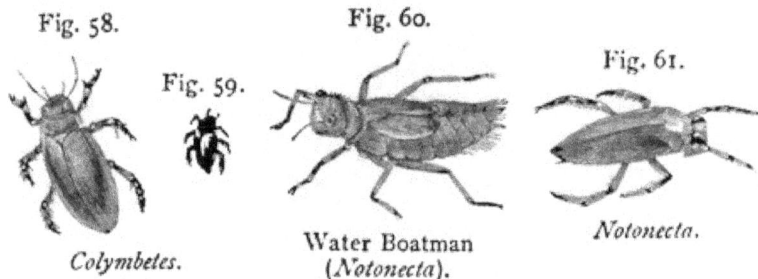

Colymbetes.

Water Boatman
(*Notonecta*).

Notonecta.

bears.   It will float in the summer sun for hours at a
time, in a complete invert attitude ; but soon and
rapidly moves when disturbed.   One or two of these
insects may be safely placed in a large tank, especially
where there are too many tadpoles, as they keep the
latter down by feeding on them.   The water boatmen
(*Notonecta*) are also hemipterous insects, deriving their
Latin generic name from their habit of swimming on

their sides or backs. They are very active, inoffensive, and interesting aquarium objects.

Fig. 62.

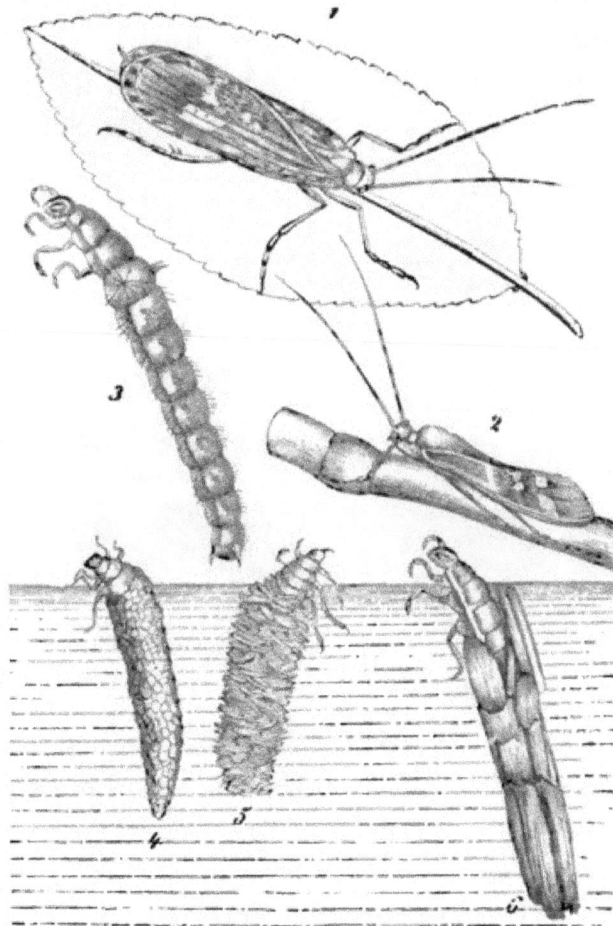

4, 5, 6. Various species of Caddis-worms (*Phryganea*).
3. Larva when taken out of case.
1 and 2. Perfectly developed insects.

The various species of caddis-worms which haunt our streams, ponds, and lakes belong to the order

Phryganeidæ. There are a great many of them, and
the larvæ of each have usually a different plan of constructing their well-known tubes, by which they may
be identified. Some select minute shells of a species
of *Planorbis* or *Pisidium;* others use grains of sand.
The genus *Limnephilus* prefers pieces of rush or other
aquatic weeds. But all of them are interesting, and

Fig. 63.

Limnephilus flavicornis.

Fig. 64.

Larvæ of *Phryganea grandis.*

Fig. 65.

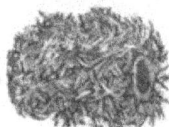

*Limnephilus
rhombicus.*

Fig. 66.

Caddis-worm pro-
jecting its head out of
tube.

seem to be perfectly aware that they are regarded as
choice and dainty bits by other larger and more active
water animals. All of them hold on to the interiors
of their frail defences by means of a series of hooks,
so that it is somewhat difficult to drag them out
forcibly. Before they pass into the quiescent state,
previous to changing into their image condition, they
protect themselves by making gratings at the ends of
their tubes. The insects into which these larvæ even-

tually pass, in many respects (notably in their having scales on their wings) resemble butterflies and moths.

The larvæ of another insect, belonging to the *Ephemeræ*, is usually very abundant in ponds. It is shown in its natural size at *b*, Fig. 67, as well as

Fig. 67.

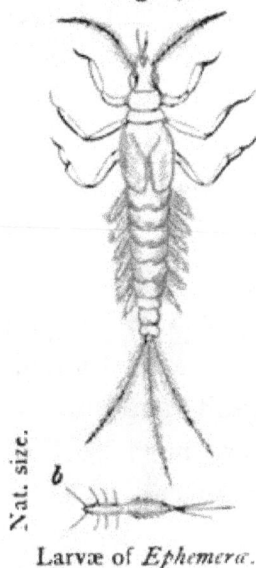

Nat. size.

b

Larvæ of *Ephemeræ*.

Fig. 68.

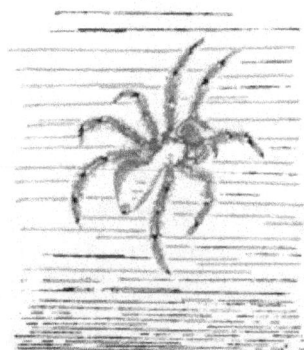

Water Spider (*Argyoneta aquatica*).

enlarged, to indicate the breathing leaflets along the sides of the body.

Nor should we forget the exceedingly interesting water spider, not only because its habits depart so extremely from those of its kind, but also on account of its prettiness and intelligence. This species (*Argyoneta aquatica*) is not uncommon. In the water it looks as if its body were covered with a film of quicksilver. This is in reality a film of air which it entangles on the surface and carries below, so as to fill the diving-bell-like nest it has spun with it, and which is air-tight. This nest is filled by successive journeys of the spider to the surface, to store up the air for subsequent breathing. These air-nests may be seen in most ponds, especially where there is an abundance of *Anacharis* and water-crowfoot leaves. There are several species of the order *Arachnidæ* which live in water, but they are usually of the kind we call "ticks." Some of them pass through very interesting metamorphoses, being parasitic on plants and insects alternately before they attain their fully developed condition.

It will be seen that the objects required for stocking ordinary fresh-water tanks are not difficult to find ; and, if proper precaution as to the habits and voracity of the different kinds be taken, there need be little fear as to ultimate and continued success. Every aquarium, large or small, ought to be as perfect an imitation of natural conditions as possible, and success usually depends on the degree to which this is carried out.

CHAPTER VIII.

THE AQUARIUM AS A NURSERY FOR THE MICROSCOPE.

To those who keep aquaria for the sake, not merely of being amused, but of learning the higher lessons which animated nature is ever so ready to teach, both fresh-water and marine parlour-aquaria may easily be converted into nurseries for microscopic research. Here may be reared with the utmost ease thousands of minute forms of life, whose ephemeral history of various conditions may be actually seen enacted upon the stage of the microscope. Human eyes can thus look down upon and witness the evolutions of these lower forms of life, just as it is possible other eyes look down upon our own terrestial career.

The fresh-water tank especially is worth supporting, even for the sake of its microscopical animals and plants alone. Mere littleness does not detract from the interest of their microscopical study, but rather throws a romantic glow about it. By the true naturalist magnitude is not taken specially into account, nor is minuteness of size regarded as in itself a sign of low organisation. It is true that most of the lower forms of life are microscopic, but this is because it is

I

an advantage to them in the peculiar conditions by which they are surrounded. Not unfrequently their small size is more than compensated for by the enormous rapidity with which individuals are produced. Many of the lowest types of vegetable life with which every tarn, pond, and stream is crowded, and which may be kept with the utmost ease ready for inspection in the aquarium, are *single-celled*. But these single cells are constantly splitting into two parts, as in the *Desmids* and the *Diatoms*, each of which becomes a new individual, and goes through the same mysterious self-division. The main difference between these peculiar objects and vegetable species of a higher organisation and greater magnitude seems to us to consist in the fact that in the former the cells are detached as fast as they are formed, whereas in the latter they adhere together, and thus produce objects of large volume. This is proved by the fact that all species of desmids and diatoms are not single cells. Not unfrequently we find them living in colonies, either for the whole or part of their lives.

Few objects are prettier than the microscopic plants we are now referring to. Seen by the naked eye their presence is perhaps only revealed by the green film covering the inside of the glass, to which aquarium keepers who are not microscopists strongly object. Desmids and diatoms often cover the stems and leaves of aquatic plants with a greenish or olive-coloured slime, such as *Hyalotheca dissiliens* (Fig. 69).

The sliminess in which many of the desmids are invested usually serves the purpose of keeping the loosely aggregated cells together, as in the species just mentioned, whose generic name is derived from this glassy sheath. But although these colonies of

Fig. 69.

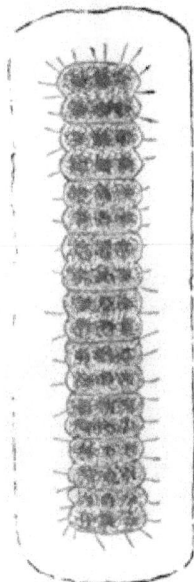

Hyalotheca dissiliens.

Fig. 70.

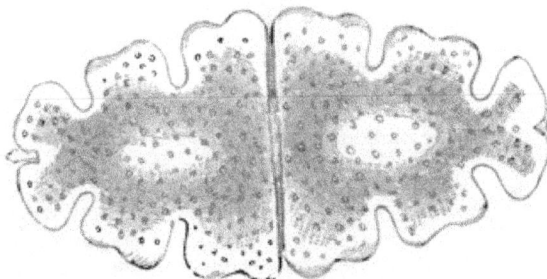

Euastrum oblongum. (Front view) × 250.

desmids are far from uncommon, they are not so abundant as the single-celled species. The latter may nearly always be seen in the act of dividing themselves into two halves—their only method of reproduction.

Among these the genera *Cosmarium, Micrasterias, Closterium*, and *Euastrum* are usually abundant, and the student may readily obtain and keep them

I 2

for examination in his glass tank.   As a rule,
however, the lower vegetable forms thrive better in

Fig. 72.

Fig. 71.

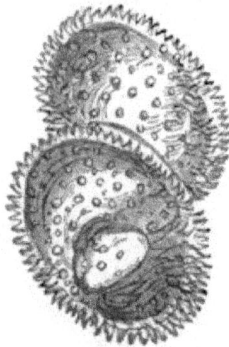

Cosmarium margaritiferum
× 250.

Fig. 73.

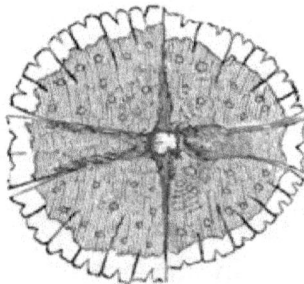

Micrasterias rotata
× 250.

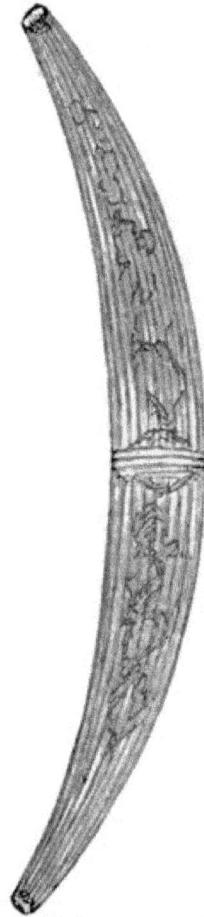

Closterium striolatum × 250.

small tanks than in large ones.   The desmids are
purely fresh-water algæ, and may readily be distinguished from the diatoms to which they are so

nearly allied, by the fact that they are always of a
light pea-green colour, whereas diatoms are usually
of a dull olive-brown. Again, diatoms are found in
fresh, brackish, and salt water alike, whereas we have
seen that the desmids are confined to fresh water.
Another most important difference is the fact that
diatoms have the power (which desmids have not) of
secreting a siliceous or glassy film on their exterior,
like that which coats the outside of straw. This glassy
film remains perfect after the diatom is dead, and is
called the "frustule." It is divided into two parts,
like the body and lid of a pill-box, so that the same
species has a very different appearance according to
the side which is looked at. These glassy cases or
frustules are indestructible, and often accumulate to
an extraordinary depth. They form a considerable
part of the black mud laid bare at low water in tidal
rivers or estuaries—they compose the greater part
of similar material in our ponds and ditches. They
will accumulate on the bottom of an old aquarium,
where they may always be obtained. A little and
patient treatment with hydrochloric, sulphuric, and
nitric acids, to get rid of the soluble organic matter,
at length displays these glassy sheaths or frustules in
all their beauty, adorned with dot and line and curve,
in the most extravagant and even luscious style of
ornamentation. We have frequently thought that
these diatom ornamentations might be studied to
some purpose by jewellers and others interested in

developing new designs. We give an illustration of
the glassy frustule of a British species of *Isthmia*,
magnified four hundred times, to prove that the

Fig. 75.

Fig. 74.

Fig. 76.

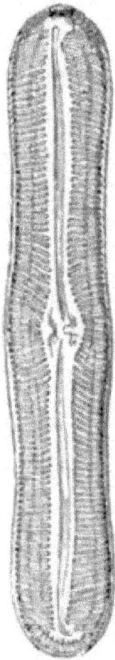

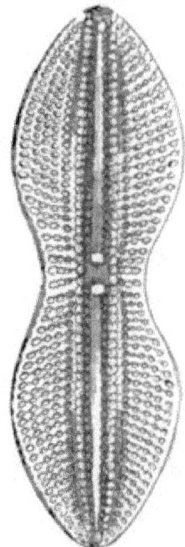

Pinnularia major.   *Pleurosigma formosum.*   *Navicula didyma.*

application of a higher microscopic power only brings
out the beauty of these forms into more prominent
relief than otherwise. In this figure the student will

also see the mode of attachment of the lower angle
of the frustule to the one beneath, by a kind of
gelatinous cushion. It is among the marine species

Fig. 78.

Fig. 77.

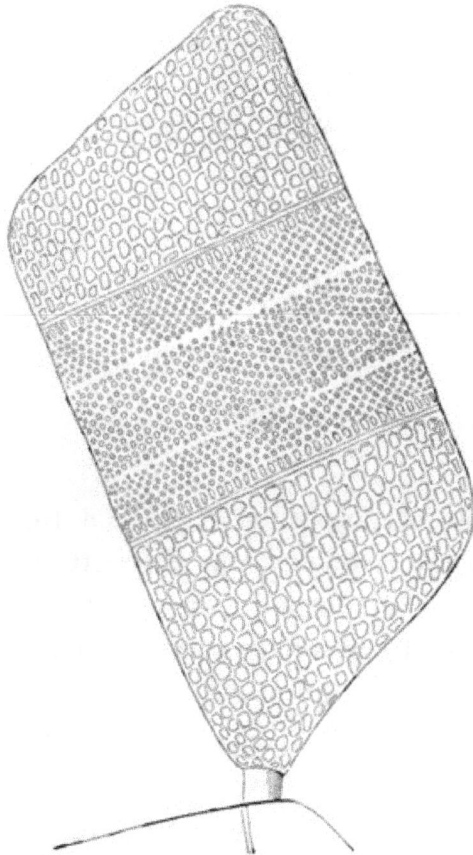

Stauroneis.

Isthmia enervis.

of diatoms, perhaps, that we are to look for those
possessing the greatest beauty, as in Figs 79 and 80
(*Cocconeis*). Among the commonest of our native

species are *Pinnularia*, *Stauroneis*, *Pleurosigma*, and
Navicula. The latter is especially abundant, and
may be seen moving about—a spectral craft, without oars or crew—amid the tangled mass of living

Fig. 79.

Fig. 80.

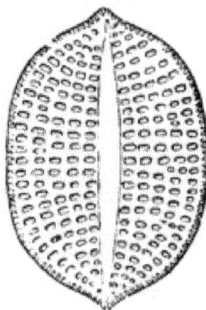

Cocconeis insida.

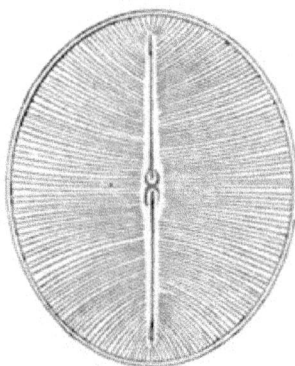

Cocconeis major.

and decomposing vegetation to be seen in a drop
of water taken from the bottom of the tank! The
boat-like outline (whence the name of *Navicula*,
and others which some of the genera bear) is then
seen to be admirably adapted to move in and out
of such interstices. It is indeed most instructive
to see these diatoms going this way and that, as
if gifted with sense, backing in and out, turning
one way and another, as if they possessed volition.
And yet we know they are merely very lowly organised cells. Division takes place by one half of the
glassy shell, with its contents, separating from the
other, just as we should take the lid off a full pill-box.

The enclosed protoplasm soon adds a new film of silica to the naked surface, and thus a new diatom is born. Many of the diatoms, like the desmids, live either part or the whole of their lives in colonies, as

Fig. 81.

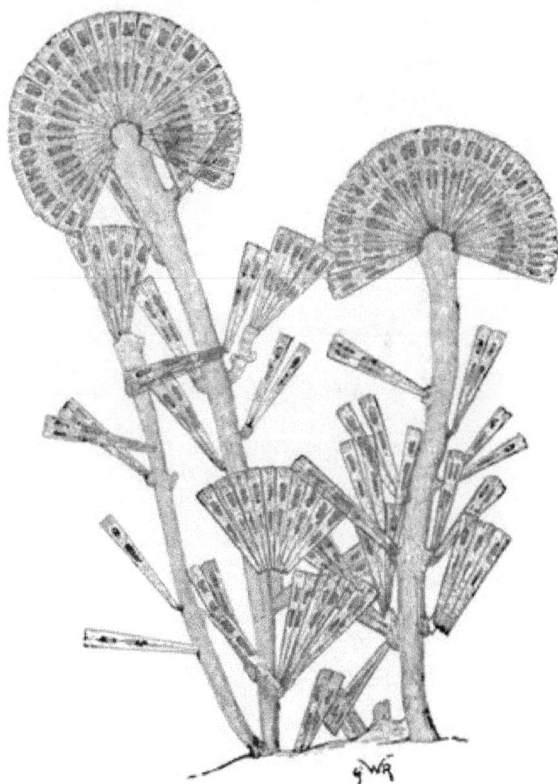

Licomophora flabellata.

for instance *Licmophora* (Fig. 81). This is a marine species, and may be found in rock-pools, where it is readily identified by its resembling a golden wool, shining like spun glass when transferred to the col-

lecting bottle.   We have magnified one of the fans to
show how closely the individual diatoms are packed on
the summit of the jelly-like or protoplasmic stalk.

Fig. 82.

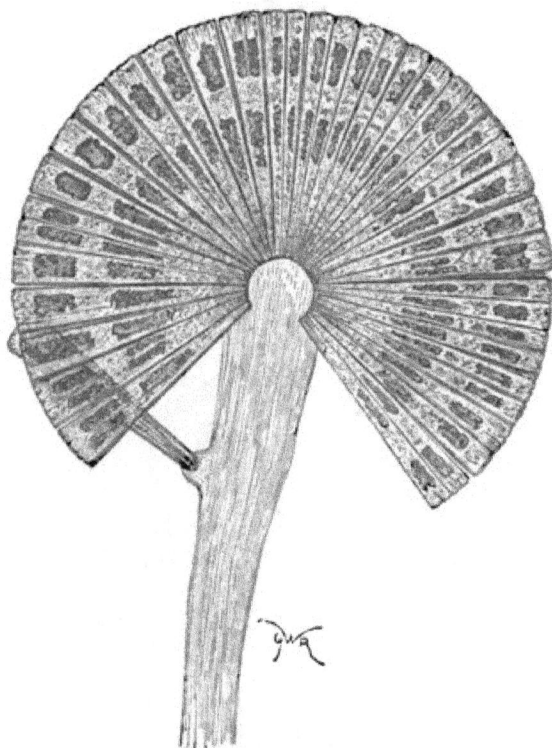

Fan of *Licmophora*.

Among the lower species of animal life which may
be transferred from the pond to the aquarium are
the *Amœbas*, rotifers, fresh-water polyzoa, hydras,
water fleas, cyclops, infusoria, &c.   The fresh-water
sponge (*Spongilla fluviatilis*), although common in

clear streams and ponds, is difficult to transfer to
the aquarium without damage. It has been effected,
however, and Mr. F. Meggy gave a description in

Fig. 83.

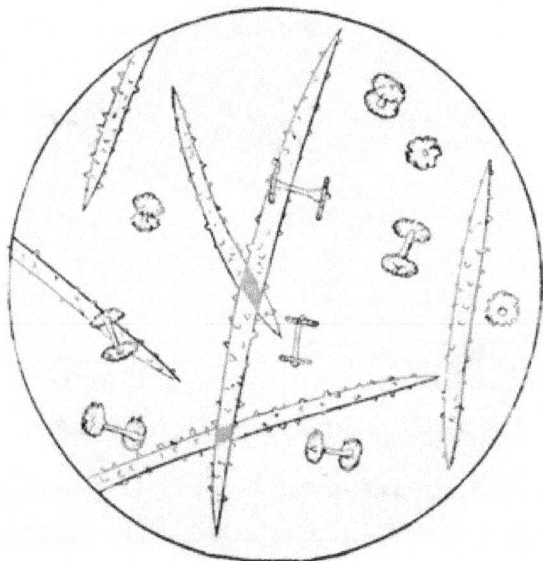

Spicules of *Spongilla fluviatilis.*

an early volume of 'Science Gossip,' of a successful
experiment in acclimatising this animal. If only
domesticated its development becomes one of the
most interesting of microscopical investigations. The
young are seen thrown off as gemmules, and the
greenish-yellow gelatinous flesh, which constitutes the
true sponge, is seen investing the spicules—or rather,
the latter are imbedded in the sarcode or flesh. The
shapes of these spicules are due to an organic crystal-
lisation, and although admirably symmetrical in form,

are no more mysterious than the crystallographic shapes of minerals. The *Amœba* is a still commoner object, which will find its way into the aquarium whether we will or no. It is a minute, gelatinous speck,

Fig 84.

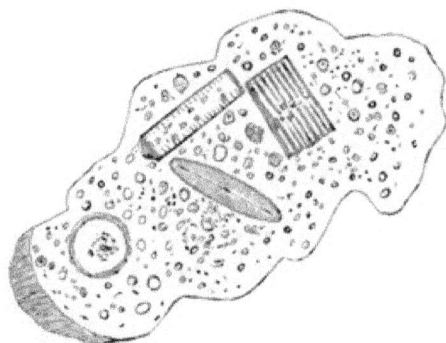

Amœba villosa, with diatoms, &c., in its interior.

possessing the power of protruding any portion of its body in any direction at will. The sarcode of all sponges is usually regarded as made up of a colony of such amœboid forms. The true *Amœba* never secretes spicules, although we may frequently see the frustules of diatoms imbedded in its mass (Fig. 84). These, however, are the objects on which it feeds, and over which it has the power of gradually pouring its own flesh until they are enclosed, and await assimilation. Then the solid parts are as gradually passed to the outside and extruded. The fresh-water hydras are most interesting aquarium objects. We have two, if not three species, one of which, *Hydra vulgaris*, is very common. They are each about one-eighth of an

inch in length, and may be seen suspending themselves from the under side of the leaves of the duckweed (*Lemna*), or the thread-like leaves of the water

Fig. 85.

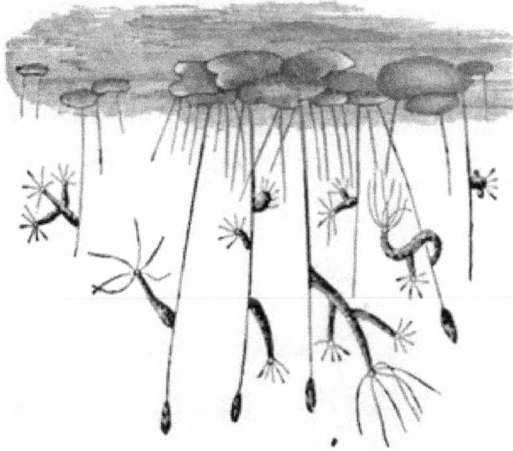

Hydra viridis.

crowfoot (Fig. 85). Few fresh-water objects have been
more studied, since the first experiments upon them
by Trembley a long time ago. When watched with
a one-inch power it is surprising what a store of
perpetual interest is afforded by them. We can then
witness the vegetable process of *budding*, in which the
lower animals are on a par with the higher vege-
tables ; we can see the sperm-cells discharged from
the tubercles, and notice the young developing
through the various stages of growth, until they
attain the parental size and shape.

Minute though these animal and vegetable forms

Fig. 86.

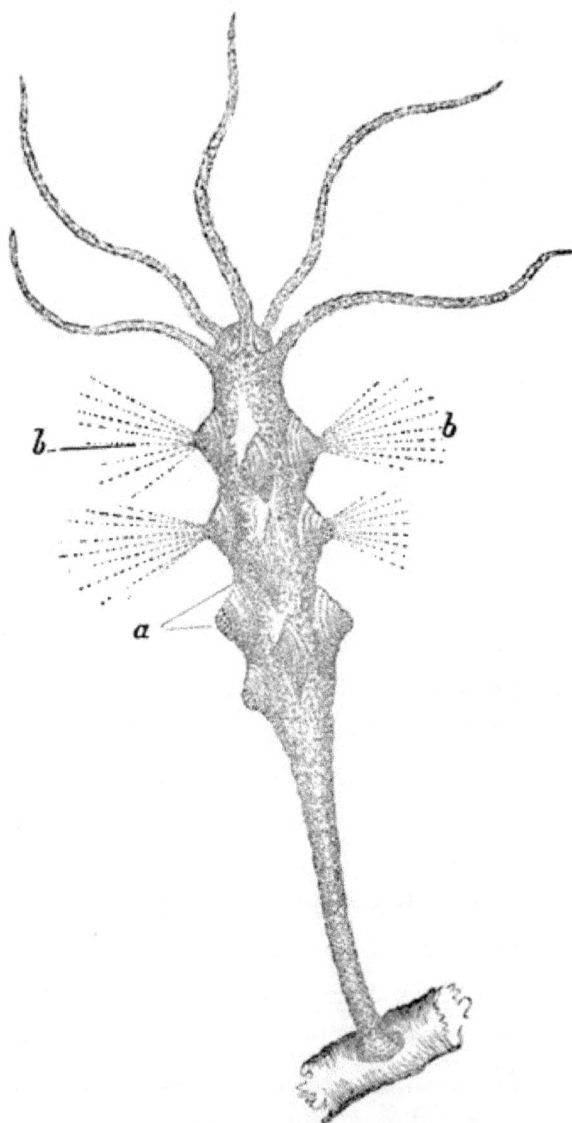

Showing *Hydra* magnified ; and at *a* prominences, which subsequently
burst as at *b*, and thence issue spermatozoa.

of life are, they are as admirably adjusted one group
to the other, as the higher animals are to feed
on and keep down the higher plants. Moreover

Fig. 87.   Fig. 88.

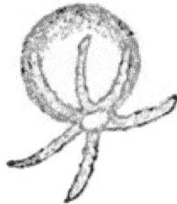

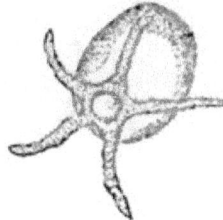

First stage of development of   Second stage of development
Hydra.   of *Hydra.*

there is the same kind of dependence on each other's
needs. The microscopic plants, such as desmids and
diatoms, give out just the oxygen which amœbas,
hydras, or rotifers require. Again, we find amongst
the lower types of animal life the same broad
differentiations which stand out so visibly among
the higher, of *carnivorous* and *herbivorous*. Death
is as much the law of the life of the microscopically
small as it is of such creatures as the tiger and
the antelope.

The rotiferal animalcules may be obtained and
kept in any abundance. Some of them, as for instance
the *Melicerta*, will be found attached to the various
small leaves of plants. These are usually deno-
minated "stalked" rotifers. All around it the *Melicerta* builds up a caddis-worm-like case, composed
of pellets of rejected or excreted food. With

a half-inch objective we may see the currents
produced by the cilia of the disk, whose rapid,
successive movement gives that impression of the

Fig. 89.

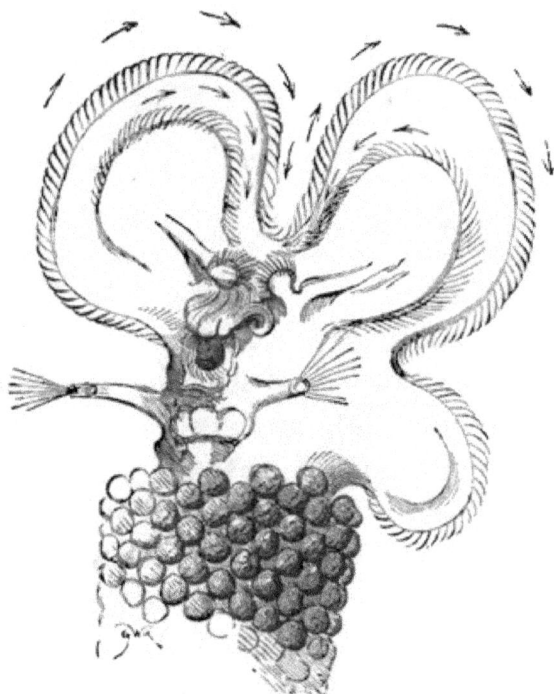

Disk of *Melicerta*, showing currents by cilia, and pelleted tube
(greatly magnified).

revolutions of a cog-wheel which has procured for
these creatures the name of *Rotifers*, or "wheel-
animalcules." Even more beautiful than the *Melicerta*
is the "crown animalcule" (*Stephanoceros Eichornii*),
Fig. 90. It may be found in many ponds and rivers,
and easily transferred to the tank. Its five arms

are beautifully feathered with cilia, and are capable of being folded up like the petals of a tulip. All the active functions of life may be seen in operation through the transparent, gelatinous envelope in which this creature is enclosed. Although fixed to one place, its maelstrom-like currents swirl into its fate-like grasp crowds of active little infusoria. The power of sudden contractility possessed by all these sessile rotifers is very remarkable. As you witness the withdrawal or collapse of the ciliated arms you are almost startled, and dart back from the tube of the microscope which is revealing the fact, as if a gun had been fired near you ! Even some of

Fig. 90.

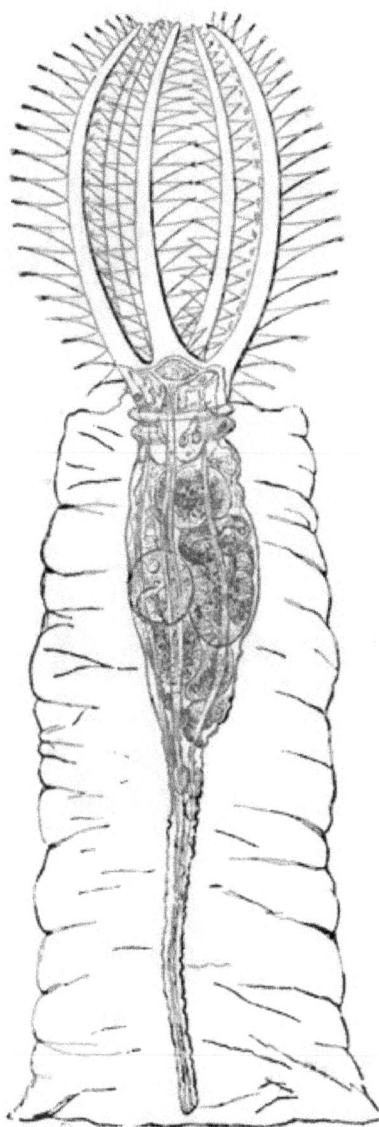

The Crown Animalcule (*Stephanoceros eichornii*).

K

the infusorians on which these rotifers feed, have their stalked and free conditions, as in *Vorticella*, *Epistylis*,

Fig. 91.

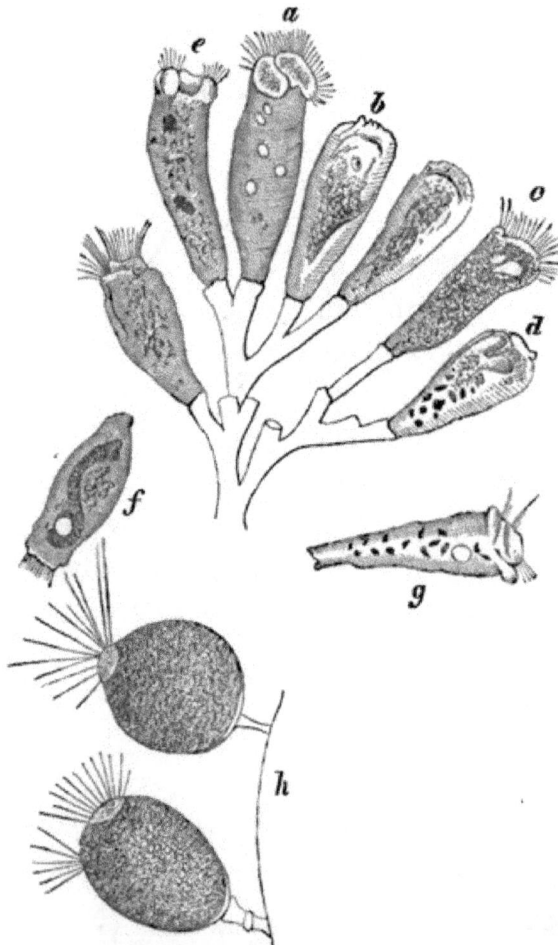

Epistylis, showing stalked individuals at *a* to *d*, and free individuals at *f* and *g*; *h*, individuals showing cilia.

&c.   The latter often grows so thickly on water fleas and cyclops, as actually to impede their movements in

the water ; and so we find them masked with a forest
of *Epistylis*, just as marine crabs often are with sertularians. The free-swimming rotifers are very active in
their habits, although we may witness them as long
as we like, quietly grazing among the microscopic
thickets of fresh-water algæ in which they most love
to disport. The play of their wheel-like crowns is
exceedingly graceful, whilst their bodies are usually
so transparent that we can see all the internal movements and processes as plainly as if they were constructed of glass. As its name implies, *Rotifer vulgaris* is very common, and the student cannot fail

Fig. 92.

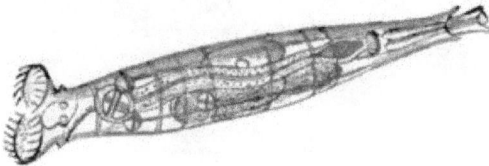

Rotifer vulgaris.

to find it in his fresh-water tank. When removed by
means of the common dipping-tube, to the slide, and
placed under the microscope, it behaves as if it had
not been transferred to new quarters, and evidently
seems to know that the enormous quantity of force
which is given off in its active habits, requires to be
replaced by a corresponding quantity of food. Hence
the great business of its life seems to be *feeding*. The
rotifer is not a gourmand ; everything seems to be fish

that is swept into its living net. You see one bit of
semi-decomposing vegetation after another descend
into its funnel-shaped gullet—now an *Amœba*, and
now an unfortunate infusorian, terminates its career
in the same trap.

Fig. 93.

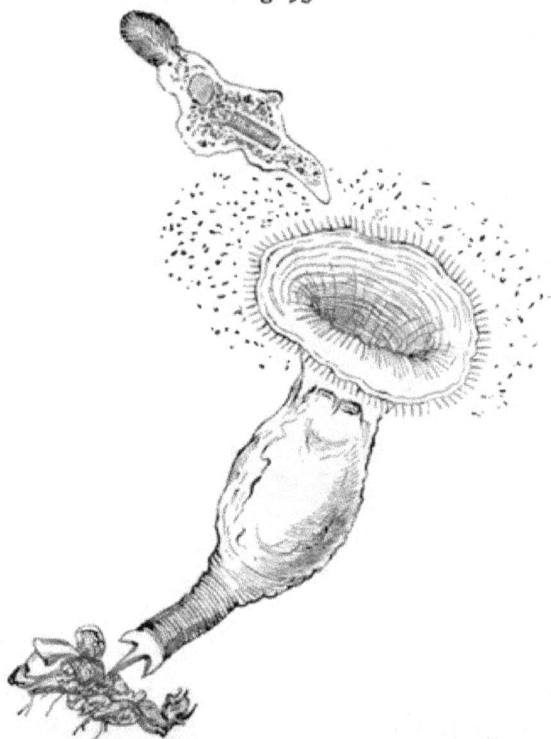

Rotifer sucking in an Amœba, by means of the current produced by
the cilia.

Not less, but infinitely more interesting when carefully observed, are the habits of another rarer group
of fresh-water animalcules called Polyzoans. They
are nearly related to sea-mats, found so plentifully

Fig. 94.

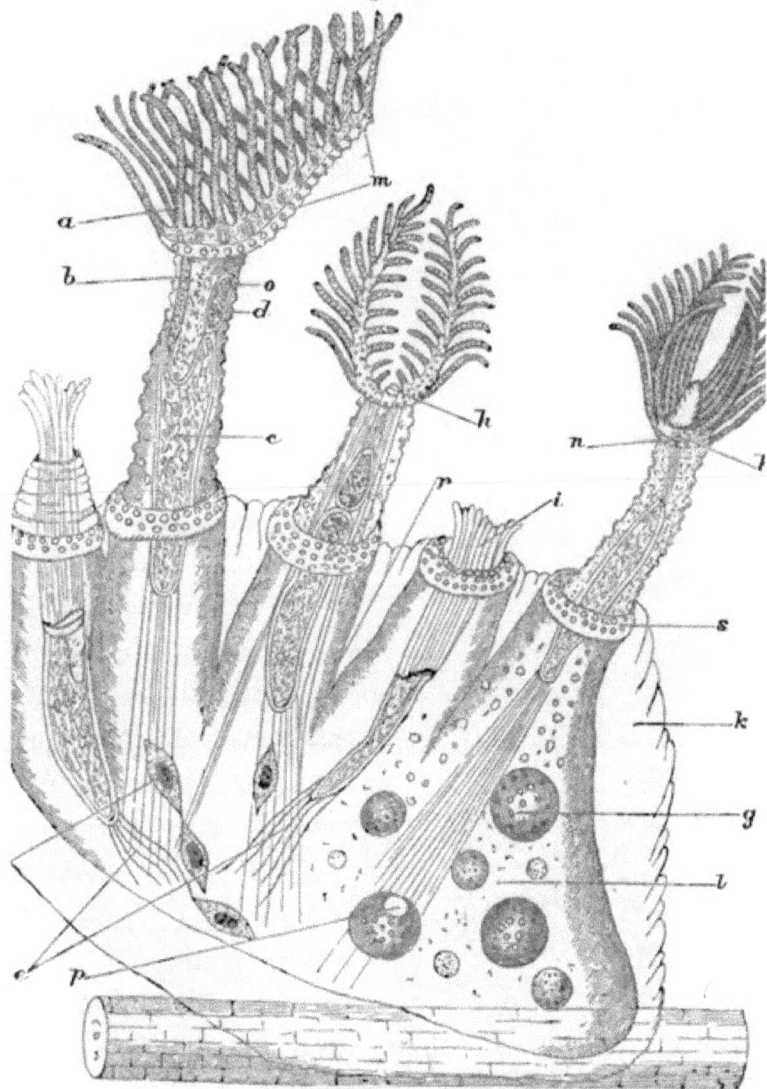

Fresh-water Polyzoon (*Lophopus crystallina*) (magnified).

on our coasts, and are somewhat out of their sphere
in *fresh* water, salt being that in which, as a

Fig. 95.

Plumatella repens investing stem of weed (magnified).

family, this group is best distributed. Nevertheless, we have several British species of polyzoa, which are not uncommon. They are well worth finding and transferring to the aquarium, for it is impossible for any other group of minute animals to afford the same degree of pleasure, so lovely are the shape and movements of these creatures. They are much more highly organised than rotifers, and consequently perform more functions. We always find them attached to the stems or thread-like leaves of water plants. Of our British genera *Lophopus* and *Plumatella* are the commonest (Figs. 94 and 95). In the figure of the former we have the region of the mouth shown at *a*, the œsophagus at *b*, the stomach at *c*, the intestines at *d*, the muscles at *e*, the mouth at *h h*, the withdrawn tentacles at *i*, the *lophophore*, or " crest-bearer," which is covered with cilia, at *m*. The same kind of horse-shoe shaped *lophophore*, or crest, surrounds the mouth of *Plumatella*.

Fig. 96.

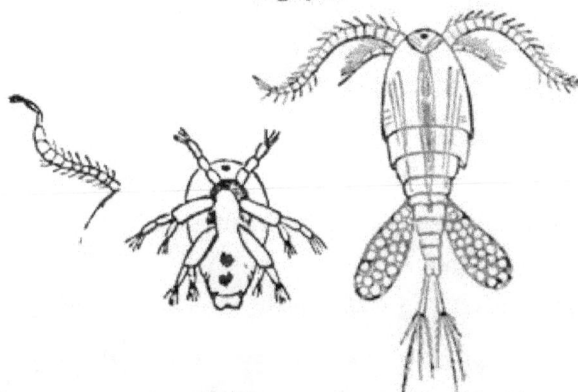

Cyclops, showing female with egg-bags ; the young, and a single-jointed antenna.

Cyclops and water fleas (*Daphnia*) ought to be encouraged in every fresh-water aquarium. Their food consists chiefly of decomposing aquatic vegetation, desmids, &c., whilst their own wonderful powers of reproduction will always people the water with living food for the higher animals. Most fishes and amphibians live to a very great extent upon these creatures. *Cyclops* (Fig. 96) is a very common aquatic object, and the female may be seen with the naked eye. So prolific is she that it is stated she would be the progenitor of four millions and a quarter of young in twelve months, if undisturbed. Her motherly devotion is seen in the way in which she swims about with her purse-like egg-bags trailing behind her. The young are ludicrous objects, with nothing of the gracefulness of their parents. They have a crab-like form, which indicates their crustacean belongings, and they move about in the most cranky, jerky manner it is possible to conceive, like the big seconds finger of a large clock.

Both Cyclops and water fleas (*Daphnia*) have such a transparent skin that we can see their internal organs, and watch them fulfilling their several functions. As is the case in most insects, the female is larger in size than the male. These creatures are a very old race, for we find them fossilised in carboniferous shales in such prodigious numbers that we feel assured they must have bred as plentifully millions of years ago as they do now, when they not unfrequently darken the water of our ponds with their countless crowds. During entire geological periods they have been the food-stock of fresh-water and marine fishes alike, and we find them adapted to all conditions of aquatic life.

Space forbids us to notice other common microscopic objects which may be readily introduced into the parlour tank, and kept ready for observation. They might thus while away many an hour of gloom and perchance of sickness or sorrow; and cannot fail to add another element of interest to aquarium keeping. The two powers which move the human world, love and hunger, seem to reign quite as predominantly

Fig. 97.

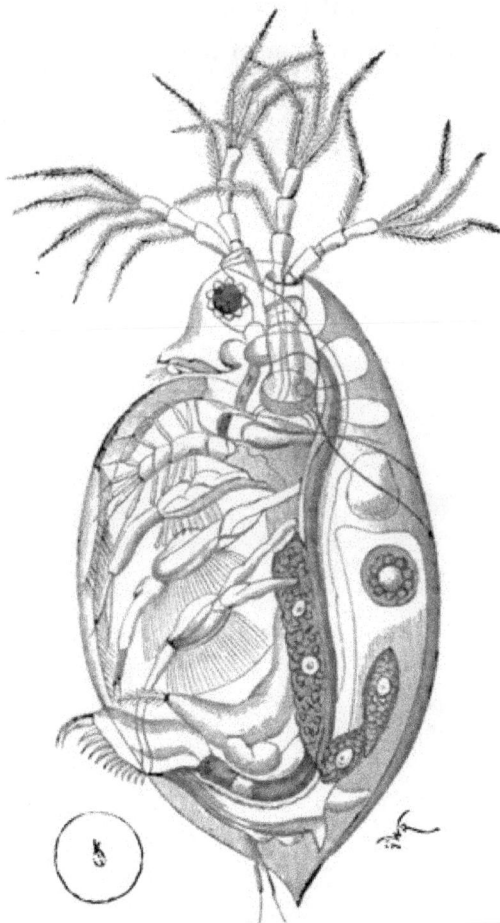

Fig. 98.

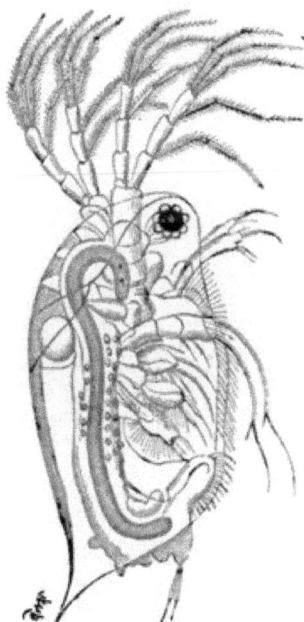

Water flea (*Daphnia pulex*) male, enlarged to corresponding proportion as Fig. 97.

Water flea (*Daphnia pulex*) female, magnified to show internal organs. The small figure enclosed within a circle shows the natural size.

among animalcules. It is not "live and let live," however, which is their rule; but "might is right!"

The weakest are constantly going to the wall, with a vigour in which there is no room left for pity, except in the soul of the beholder! May not that be the case with ourselves, and the Great Power whose wisdom has called us and them alike into existence, to fulfil some inscrutable but all-wise purpose of his own?

CHAPTER IX.

MARINE AQUARIA FOR ROOMS. SEA-WEEDS FOR DITTO.

THERE are undoubtedly greater difficulties attending the healthy maintenance of small marine tanks in rooms than fresh-water ones; but some of these difficulties vanish before a little common-sense treatment and knowledge of the habits of the creatures we endeavour to keep. As a rule the marine aquaria kept in houses are on too small a scale, and there is the unconquerable tendency on the part of their owners of putting as many objects in them as they possibly can. There is really no reason why small marine tanks should not be kept in a good condition for years, provided they are attended to, say as we should attend to poultry, rabbits, guinea-pigs, canaries, or any other pets we are rearing under semi-artificial conditions. If these are neglected, everybody knows the consequences ; and we cannot expect that healthy aquaria can be maintained and neglected at the same time. Indeed, considering how altered are the circumstances under which they live, the wonder is that the inhabitants of aquaria give so little trouble.

In constructing small tanks to hold sea water the

same cement may be used as is mentioned when
speaking of fresh-water aquaria. The glass sides
may be secured by using Collin's patent elastic
and marine glue. Rockwork is more indispensable
in a tank of this kind than it would be in fresh water,
where it is often in the way, rather than otherwise.
It ought to be built up in the middle, unless there are
reasons to the contrary. Pumice-stone, pieces of mica

Fig. 99.

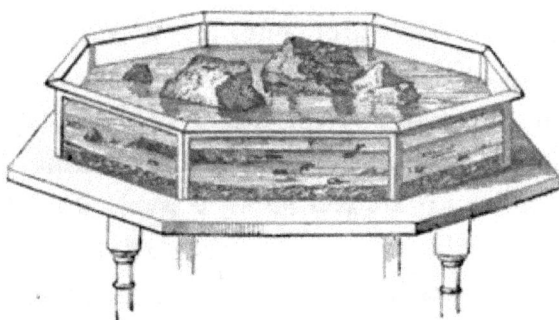

Octagonal Marine Table Tank.

schist, the slag obtainable from brick-kilns, and oyster
shells covered with serpula, are the best kinds of material, inasmuch as they do not give off anything that
will affect the water. The rough surfaces of the pumice-
stone and mica soon get greened over with minute and
semi-developed algæ, and then look very pretty. The
rockwork ought to be built in arches and caverns, so
that the inhabitants may find shelter from the glare
of the too strong light, and cool spots where they can
repose. Nearly all the crustacea shun the light when

they are about to moult, and such caverns and grottos are to them very necessary. Very fine, well washed sea or river sand should be strewn on the bottom of the marine tank, and the depth ought to be adjusted to the creatures it is intended to keep. If marine worms, such as *Sabella* and *Terebella*, and even some crustacea, it should be of considerable thickness; but always with fine shingle mixed with it, for these creatures love to burrow in it. Shrimps take great pleasure in dusting themselves with the fine sand, descending into it and throwing it about them much after the manner of birds.

As only few kinds of sea-weeds can be successfully grown in small aquaria, it follows that aeration or oxydization of the water has to be compassed by artificial means. The sea-lettuce (*Ulva latissima*), Fig. 100, is believed to give off more oxygen than any other marine plant, and is well acted upon by the sun in clear, shallow water, insomuch that we may often see its bright green surface silvered over with minute bubbles of oxygen. The best way to transfer sea-weeds is to find small specimens rooted to pebbles or rock, and bring them as they are thus found, taking care not to break or bruise them, otherwise they will slough. *Cladophora* is a sea-weed which grows readily, and when its thread-like fronds are expanded, it has a very pretty appearance. If the student is not perplexed by a little extra trouble, there will be little difficulty in growing other sea-weeds;

but he must remember they are by no means so easily reared as fresh-water plants, and when they

Fig. 100.

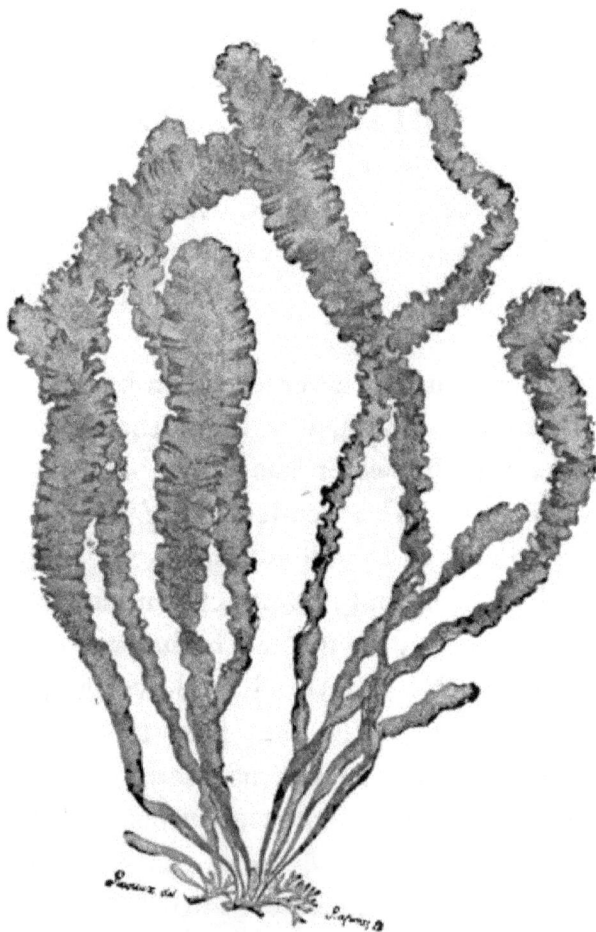

Sea-Lettuce (*Ulva latissima*).

sicken and die they taint the water, and give off a noisome smell. We have seen aquaria, however,

in which both red and green sea-weeds were growing
and looking as bright as a garden parterre. To
bring these somewhat capricious plants up to this
pitch is well worth the trouble. In this way it is not

Fig. 101.

Callithamnion.

impossible to have a marine garden of such lowly
organized "flowers of the sea," as they have been
somewhat sentimentally termed. The red sea-weeds
should be planted in the darker places, for they shun
the light more than the green and the olive-coloured,
most of which, on the contrary, seek it. The best
sea-weeds are the smallest. The student ought by
no means to attempt the larger wracks, as they give

off a good deal of mucus when sickly, and poison the
water. The following species are the prettiest, and offer

Fig. 102.

Delesseria sanguinea.

Fig. 103.

Plocamium plumosum.

a certain variety of colour and tint, which make an
aquarium look very pretty: *Ulva, Callithamnion,*

Cladophora rupestris or *arcta, Bryopsis plumosa, Grif-
fithsia setacea, Corallina officinalis, Enteromorpha intestinalis, Ceramium rubrum, Rhodymenia palmata,
Delesseria sanguinea, Padina pavonia, Plocamium
coccineum,* &c. *Corallina* requires a good deal of
lime, for it will cover the rock on which it is growing

Fig. 104.

Padina pavonia.

with a film of pinkish-white limey matter. If the
tanks be large enough the small wrack (*Fucus canaliculatus*) and the carageen or "Irish moss" (*Chondrus
crispus*) may be grown. The latter takes to its ground
very readily, and is really a very pretty plant, liable
to assume different tints according to its surroundings.
The fructification of the small wrack is a very interesting performance, and may be studied in the aquarium
if this species can be induced to grow. We have seen

L

another not uncommon sea-weed (*Halidrys siliquosa*) grown in tolerably large marine tanks, and looking both healthy and attractive. The *Zostera marina*— a true flowering plant, and not a sea-weed—may even be grown where the tank is large enough. This is undoubtedly a most useful plant for giving off oxygen. Mr. Shirley Hibberd recommends *Codium tomentosum*,

Fig. 106.

Fig. 105.

Corallina officinalis

Plocamium coccineum.

not only on account of its growing easily, but because it is a favourite food-plant for many mollusca, &c. For it should be remembered that if certain animals, hereafter to be mentioned, are introduced, they will feed on many of the sea-weeds ; so that the latter have to serve the double purpose of aeration and

provision. Indeed, there can be little doubt that
one of the "stock" kinds of food for many marine
creatures consists of the zoospores, &c., of sea-weeds.

Fig. 107.

Rhodymenia palmata.

Nitophyllum is another genus of beautiful green sea-
weeds, readily obtained in true rock-pools. Autumn is
the best time for introducing sea-weeds, as the spores

are then given off, and next spring many of these
may sprout into a healthy vegetation.*

It should be remembered, however, that in a parlour
tank, small or large, there will not be sufficient aeration

Fig. 108.

" Irish Moss " (*Chondrus crispus*).

produced by sea-weeds, unless the number of animals
is very few. To assist in this most important end, a
fountain, constructed like that in Fig. 1, may be advantageously used. Some people content themselves

* In the Crystal Palace and other aquaria, some species of sea-
weeds have spontaneously made their appearance from spores. Mr.
Saville-Kent thinks that the larger sea-weeds might be grown where
the rapid circulation of the water produces a strong current.

with using a syringe every now and then, just to mechanically entangle air in the injected water. There is no harm whatever in this method, but we prefer the above fountain. In the huge tanks of our public aquaria the sea water is aerated entirely by various mechanical processes, which will be shortly described.

Fig. 109.

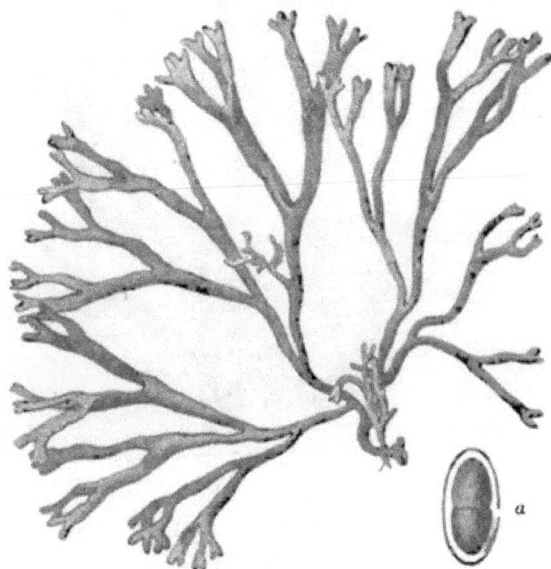

Small Wrack (*Fucus canaliculatus*). *a*, Spore-case.

No paint should be used, in the interior, at any rate, of any aquarium; and if it can be avoided on the outside it is all the better, as the presence of paint seems to cause annoyance and sickness to all kinds of animals. When the sea water becomes putrid it is for want of oxygenation; and a little artificial oxygen gas, forced into it from a bladder, to the mouth of which a clean, long tobacco pipe has been fastened, so that it can reach the bottom, might, under some conditions, soon put it to rights.

Fig. 110.

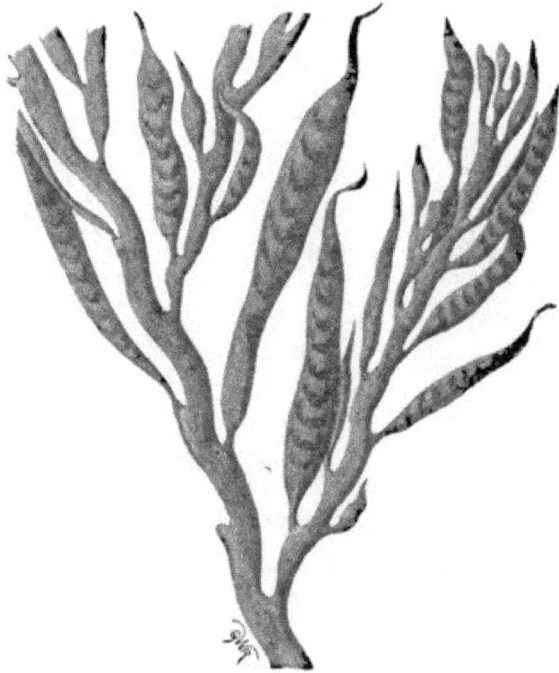

Halidrys siliquosa (nat. size).

A most useful and beautiful animal is the Ormer (*Haliotis tuberculata*), or "Venus's ear," as it is some-times called. This univalve is common along the shores of the Channel Islands. In the aquarium it is as good as a natural scrubbing-brush for keeping the sides, &c., clean; whilst it is equally useful in

devouring decaying vegetation. The common periwinkle (*Littorina littorea*) is another good scavenger,

Fig. 111.

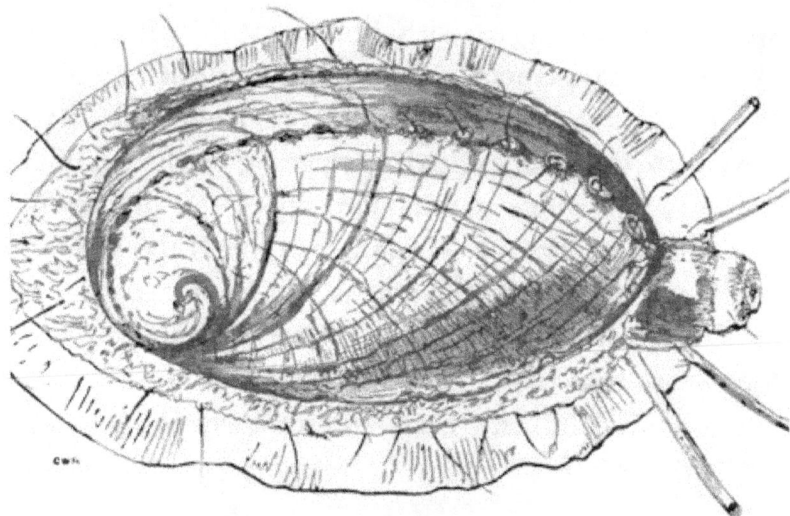

Ormer (*Haliotis tuberculata*).

Fig. 112.

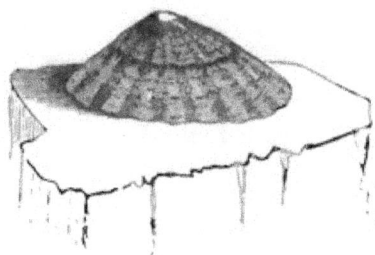

Limpet (*Patella vulgaris*).

but a voracious feeder. The limpet (*Patella vulgaris*) does its work more slowly, but effectively, and

cleans the glass and rocky surfaces. Like the periwinkle, it will often crawl above the level of the water.

Mr. W. R. Hughes recommends Mr. Edwards's
plan of having a parlour tank with a sloped back, and
we thoroughly agree with him that this is by far
the best arrangement. It has the merit of enabling
the animals to adjust themselves from the shal-
lowest to the deepest parts according as their habits
require ; and offers a minimum space for the reception of the rays of sunlight. Over this sloping
back, or bottom, rockwork is to be arranged, as
already described. The sloping bottom may be perforated, so that the tank is divided equally into a dark
and a light chamber. Such an arrangement sets up a

Fig. 113.

Section of Slope-back Tank, showing Dark Chamber.

slow, natural circulation, whilst it prevents the undue
development of the spores of algæ. Another means
of promoting circulation is that of having syphons,
which may be made to gently pour in the tank a
fine stream of water from a jar placed above. If
the current is made to flow a few inches through the

air, it will mechanically entangle some of it and thus convey it to where it is required. But the ingenuity of a really careful aquarium-keeper will soon suggest to him various simple means of this kind, by which additional aeration can be obtained if required. The water referred to is of course taken from the tank into which it is made to run. In the volume of 'Science Gossip,' for 1870, the following contrivance is mentioned : " I had been in the habit of drawing off the water to a certain extent daily, but in time it became apparent, as my interest began to flag (as it sometimes will) that this was a somewhat fatiguing process, and therefore one that was likely to be forgotten for a day, thus greatly deteriorating the condition of the water. In order, therefore, to make the tidal arrangement as self-acting as possible, I added the following : At the level of high-water mark I made a hole in the slate back of my tank ; through this I inserted a glass tube (which may be made to fit completely water-tight by placing over it a piece of indiarubber tubing). This tube *inside* is bent down like a syphon to the level of low water in the tank ; on the *outside* it communicates with the jar receiving the off-water. Now it will be seen that as soon as the water reaches the level of the tube passing through the back of the tank, it will flow over into the outer tube communicating with the jar, and this tube, by acting as a syphon, will draw off all the water to the lower water level, which in my aquarium is about 2 gallons. This is, of course, a great saving of trouble, as with only once attending to it, the water is made to rise to the highest level, and, instead of getting heated in the aquarium by remaining there, it is drawn off into the darkness again. I think that with this arrangement it is almost impossible that the water should get foul, except by the grossest neglect." Whenever a *film* or scum is seen to settle on the surface of the water, it should be cleared off, either by laying sheets of blotting paper upon it, as you would remove dust, or by drawing off the surface water by means of a syphon. Such a film hinders the free passage of air to the water, and Mr. Lloyd has shown that water has a stronger affinity for the oxygen of the atmosphere it is in contact with than its nitrogen.

The regulation of the light and temperature of marine tanks is quite as important a task as in the management of fresh-water aquaria, if not more so.

Their temperature should always be *low*, so that the water feels cold to the hand. It should, however, never fall below forty degrees Fahrenheit. If the tank be so placed that it is difficult to prevent the heat unduly warming it, a small piece of ice may be placed in the water to replace that lost by evaporation. Light is perhaps even more to be studied than temperature, as it stimulates the growth of the well-known green confervæ which are such a source of trouble to the uninitiated. If the light and temperature are both too powerful, we cannot expect other than the utmost disorder and zoological anarchy. Mr. W. R. Hughes, speaking on this point, says, "It is certain that *light* is the primary question to be considered in relation to aquaria. The presence of an uninterrupted volume, combined with a high rate of temperature, may in a few days convert an aquarium, which was in an efficient condition of health and beauty of the inhabitants, into a decomposing mass." Moreover, so auxiliary is the effect of light or heat respectively, that it is necessary to reduce the light admitted to a tank when the temperature is rising. Fifty degrees, or from that to fifty-five is about the highest heat which should be allowed; and the light should be toned down as the water reaches the higher point. Mr. Lloyd insists with great emphasis upon the necessity of keeping the temperature low, and this, he contends, can only be effectively done by having a store of water several times the bulk of that in the show tank.

It is evident that, do what we will, some of the sea water cannot fail to be evaporated during the process of fountain or syphon aeration; or even when the surface only is exposed to the air. But it should be remembered that what is thus lost by evaporation is only the *fresh* part of the water. The salts are left behind—we cannot evaporate them; and so they tend to render the remaining water all the salter and denser. Hence there is an immediate necessity for restoring the quantity of water lost, by adding a supply of *fresh* water. Distilled water is the best to replace the evaporated, as then we know no germs or other interfering agents are likely to be introduced.

In constructing a marine aquarium there is no doubt whatever that pure sea water is the best. This can now be obtained through dealers, or the student may obtain it for himself, always taking proper precaution that the barrel or other means of conveying the salt water, contains nothing that would render it obnoxious. Always use the water as soon after getting it as possible, and do not allow it to remain in wooden vessels which may discolour it. When placed in the aquarium, add the *Ulvæ, Cladophora,* or other useful, oxygen-yielding sea-weeds ; and let them grow before putting in any animals. When you think the water is fit, place in a few sea-anemones, adding them one at a time. It is much better to proceed slowly, than to be in a hurry, and have all the work to do over again. Mr. Gosse showed that it was possible to *manufacture* sea water, and gave the following formula for it :

Common table salt	3½ oz. avoir.
Epsom salts	¼ ,,
Chloride of magnesium	200 grains
Chloride of potassium	40 ,, troy

which should be added to a little less than 1 gallon of distilled water. This artificial "sea-salt" is now specially prepared for aquarium keepers, and may be purchased at the natural history dealers. We confess, however, to a disinclination to its use, although we have seen some healthy and beautiful marine tanks, full of life and vigour, supported by water prepared in this manner. At Berlin and Hanover, the marine tanks have their sea water manufactured artificially. The best way to use it, after mixing carefully in an earthen jar, is to allow it to stand in a quiet place, with a few sea-weeds thrown into it, and afterwards expose it to the sunlight. Strangely enough, spores will then develop in it, having come from the atmosphere ; or been given off by the sea-weeds. Owing, however, to the conveniences which railways and railway excursions now afford, almost every intending aquarium keeper can get his own sea water from the sea itself.

Sea water can easily be made to indicate the relative saltness or otherwise by using the " specific gravity beads," sold by London dealers. There are two kinds of them, both in the shape of pretty thin glass balls. One of these floats when the water is of the right strength, and the other sinks. Directly the *floating* ball begins to sink it indicates that the water is weak in saline matter ; whilst if the *sinking* bead rises, it is time to add a little fresh water until it falls again. Both these kinds of balls, or floats, are used in small tanks, and they are usually differently coloured, so as to be soon recognised. A good hydrometer, however, is much easier to use, and is altogether better.

Fig. 114.

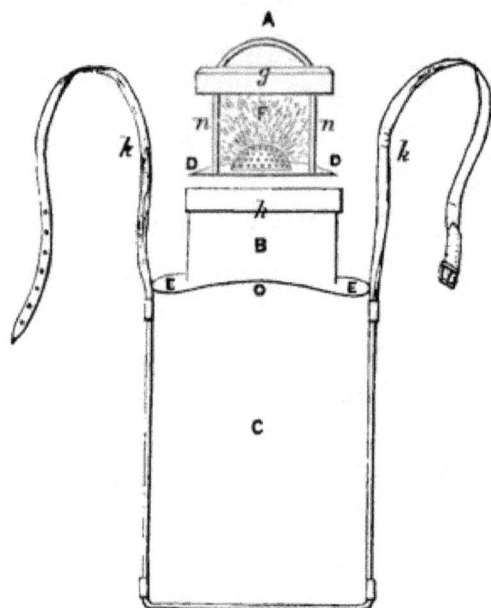

Self-acting Air-can for Aquatic Animals.

Fig. 115.

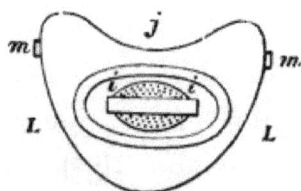

Section of ditto.

Undoubtedly one of the charms of marine aquaria consists in the personal collection of the objects.

Most people have now their annual seaside "run," and it is a poor place indeed if it have no rock-pools and no marine animals and plants there to be captured, stowed away, and brought home as souvenirs of pleasant summer rambles. We have known a good many instances, however, in which the animals thus collected have been lost for want of a proper means of keeping them until they could be transferred to the tank ; and we therefore give the accompanying sketch of Mr. A. J. R. Sclater's "self-acting air-can," as one of the best contrivances we have hitherto seen for keeping aquatic objects, whether marine or fresh water. The following is the explanation of its structure :

A is the cover of the can ; B the socket to fit the cover ; C the body of the can ; D D plate with perforated raised zinc, showing the water forcing itself into the chamber F, and then going back into the can again ; *g* the top of the cover, which rests on line *h*, sinking down one inch below upper rim, so that water forcing its way through the upper cover plate also goes back into the can again through the holes, as marked in cover on the under drawing *i i*; *j* shows the hollow, and how it is fastened to the left side by a leathern strap *k k* passing over the head to right shoulder. It can thus be worn when riding or walking. L L is the cover of the can closed down into the socket ; *m m* tin loops to pass the leathern strap through ; *n n* pillars to hold the under plate of cover ;

the plate not to go below o.   Of course it is always best to transfer any captured aquatic objects to their new habitats as quickly as possible. Keeping them in unnatural conditions is cruel, and no true naturalist will inflict pain on the humblest creature if he can possibly avoid it.

CHAPTER X.

OUR PUBLIC AQUARIA.

THE establishment of large public aquaria in English cities and towns is the best evidence we could desire of the progress of zoology. There can be little doubt that these important institutions will react favourably on scientific education by familiarising people with objects they were previously only acquainted with in books, and also by stimulating young minds to their further study. Their value to natural history cannot be overstated, for they afford means of observation which never existed before, both to study the habits and the embryological development of marine animals.

To Mr. W. A. Lloyd belongs the merit of successfully carrying out the idea of large public aquaria to their present issue. No other naturalist has enjoyed such a long and specialised experience in their construction and management, either at home or abroad. The continued success of the Crystal Palace Aquarium —which may be called the *first* public one of any magnitude, those at the London and Dublin Zoological Gardens and elsewhere being on a much smaller scale—undoubtedly encouraged the construction of

M

aquaria at Brighton, Manchester, &c.   A full and
detailed description of that at the Crystal Palace will
be found in the excellent 'Handbook' which Mr.
Lloyd has written.*

In large aquaria it would be utterly impossible
to sufficiently aerate the water in the huge tanks by
means of algæ.   The quantity required to be grown
to oxygenate the water sufficient for a very few fish
to be healthily kept would be so great that it would
almost fill the tanks.   Moreover, in these show tanks
it is necessary there should be as little as possible to
obstruct the observation for which they are constructed.   Even if sufficient aeration could be produced by the presence of sea-weeds, that alone would
not represent the actual marine conditions by which
fish and other animals are naturally surrounded.
The seething, restless condition of the ocean—with its
huge volume of water moved by tides, currents, and
storms, so that the waves raised by the latter are
always entangling quantities of atmospheric air all
over its broad surface—is best imitated in large
aquaria by the circulation which is constantly hurrying
masses of water from one place to another, and always

* In 1861 Barnum had two white whales captured for him at the
mouth of the St. Lawrence, and conveyed alive to his museum at New
York, where they were exhibited in large tanks constructed for the
purpose. Other tanks were shortly afterwards constructed by him, in
which sharks, porpoises, "angel" fish, &c., were shown.   These
animals were kept alive by a stream of salt water from high tide. This
was the first rude attempt at aquaria in America.

presenting new surfaces to the oxygenating influences of the air. In fresh-water aquaria we need this mechanical aeration and circulation in the same degree, perhaps, although we there imitate the quiet condition of our still ponds and tarns. In our small parlour aquaria, constructed to maintain marine creatures, we also require its assistance, such as may be produced by the means already described; but it should be remembered that here we are only imitating the natural conditions of rock-pools. On the other hand, in the huge tanks seen in every large public aquarium, it is sought to imitate the conditions of the open sea.

A most important mechanical contrivance for aerating aquaria was invented by the late Mr. G. Hurwood, of Ipswich, in 1859. It consisted of an arrangement by which "the pressure of a stream of fresh water, such as exists in the pipes of waterworks in towns, or such as can be got from a high cistern already existing in a dwelling house, may be employed to compress air, which compressed air in its turn forces a current of sea water into an aquarium." This contrivance was first successfully adopted on a large scale at that erected in the garden of the Acclimatisation Society of Paris, in 1859, shortly after Mr. Hurwood invented it. It was afterwards partly applied by Mr. Lloyd to the Hamburg Aquarium, of which he then had the charge. The sea water in the Hamburg institution was "circulated partly by a water-pressure engine set in motion by the town waterworks, which drives a pair of water-pumps (instead of compressed air, as was done at Paris), and partly by a steam-engine which drives two other pumps."

Mr. Lloyd's plan of keeping a large underground, dark reservoir for storage purposes, into which the water runs from the tanks after circulating through-out, and from which it starts again on its circulatory round, has been markedly successful. The water is bright and sparkling, and its temperature is thus always easily kept at from fifty to sixty degrees. The aggregate contents of the tanks at the Crystal Palace is only one-fifth of the contents of the reservoir. This readily enables the manager to at once empty any tank into it, should it get wrong, and the slight admixture of the turbid water would be unable to affect the good condition of the general volume. No animals are kept in the reservoir; the main aeration is produced by the mechanical agitation of circula-tion, and the constant injection of sprays of salt water entangling air into each tank. This is constantly going on, night and day, duplicate steam-engines and boilers being employed, in case of any accident occur-ring to one of them. The stoppage of this me-chanical circulation for some hours is attended by distressing symptoms, and Mr. Lloyd remarks that "the creatures in the tanks, and especially in the taller tanks, must be considered, to some extent, in the light of persons in a diving-bell, whose existence depends on continuous pumping and injection of air." The sea water issues from the pumps at the rate of from five to seven thousand gallons an hour, passing into the two largest tanks first. Thence it runs north and south, passing into and feeding all the rest. The sea water is *unchanged*, except about two per cent. of the whole, which is added to compensate for leakages, and one-half per cent. of *fresh* water, to supply the loss by evaporation.

Mr. Lloyd's method of employing large storage reservoirs has lately been attacked by Mr. Saville-Kent, in a paper read before the Society of Arts in March, 1876, chiefly on the ground of its great expense, and also that so extensive a store of salt water is not required.* At the Brighton Aquarium the tanks are aerated by jets of *air*, injected into the water in its simple form; so that the mechanical arrangements are quite different from those at the Crystal Palace. Indeed, we may regard these two systems as being more or less on their trial before the world. Each has its merits, but that in practice at the Crystal Palace has been longer in existence, and has never shown signs of failure. At Brighton, moreover, the

---

* Mr. Lloyd, in two contributions, one in the 'Journal of the Society of Arts,' for March 24, 1876, and one in the 'Popular Science Review,' for July 1, 1876, has sought to demonstrate by figures that the money capital of public aquaria cannot be more profitably spent than in large reservoirs of from five to even ten times the aggregate capacity of the show tanks.

adjacency of the aquarium to the sea somewhat obviates the necessity for a large storage reservoir. The sea itself may be regarded as partly acting in that capacity, for all the salt water in the tanks is pumped directly from it into underground reservoirs, capable of holding half a million gallons of water. It takes about ten hours to fill these reservoirs. The circulation of the sea water in the tanks is carried out by means of compressed air, which is supplied to their *lower* parts. One supposed advantage in this method of directly injecting air into the bottom of each tank is that it will ascend through the entire volume of water, and will, moreover, first come into contact with any organic substances lying at the bottom, which require oxydisation. Another assigned reason for employing this system of aeration and circulation is its greater cheapness. This, however, Mr. Lloyd emphatically denies ; even holding that it is eventually more expensive. We may expect, however, that many of the public aquaria founded at seaside towns and resorts will be established on the Brighton pattern, owing to the advantages which contiguity to the sea confers, by enabling the manager to pump in the sea water direct, and as often as may be required. It has been found that the salt water near the shore is quite fit for aquarium use, especially if it be kept a time in the reservoirs, so that any mechanical sediments, &c., may subside. Another assumed advantage of the Brighton system of direct

aeration by jets of atmospheric air is that it allows
of each tank to be treated independently of the rest,
if required. The Scarborough Aquarium is aerated
on the same plan as that at Brighton. Mr. Lloyd's
method of injecting sprays of water which carry down
fine air-bubbles to the bottom of the tanks, and then
allows them to be distributed through the volume of
water, however, has the support of all aquarium naturalists. Undoubtedly the expense of constructing
storage reservoirs commensurate with the size of the
largest public aquaria is very great, if carried out in
the proportion of the method employed at the Crystal
Palace ; but if it be the means of preserving both
sea water and animals in a continually healthy
state, it cannot be deemed too great. Mr. Saville-
Kent states that as the show tanks of the aquarium
at Great Yarmouth are to hold 200,000 gallons of
water, "it would be necessary, in order to maintain
the same ratio in the reservoirs as obtained at Westminster and the Crystal Palace, to construct reservoirs
large enough to hold no less a qantity than 1,000,000
gallons." The Westminster Aquarium was constructed under the direction of Mr. Lloyd, and, as
already stated, it possesses show tanks of the capacity
of 150,000 gallons. To keep up the healthy circulation of this water, after the manner described as
being in practice at Sydenham, there is storage
accommodation for 600,000 gallons, in underground
reservoirs which resemble three railway tunnels placed

side by side, and which occupy the entire space
underneath the grand promenade.

The Manchester Aquarium is a good example of
what can be effected in constructing and maintaining
marine animals under purely artificial conditions. The
smoky atmosphere of that town is proverbial, and
perhaps the least said about *directly* aerating the
tanks with it, as at Brighton, the better! From our
experience, we should say it would be an additional
element of acclimatisation for marine animals to enjoy
Manchester air, seeing that even human beings get as
far away from it as they can! This aquarium was
for a long time under the direction of Mr. Saville-
Kent, who states that it has been most successfully
maintained with storage reservoirs constructed to
hold a supply of water only equal to that contained
in the show tanks. He further declares that, " practically it has been kept in the highest state of efficiency,
with the water clear, and an abundant supply of fish
of the largest size, with less than one-half of this full
complement in the reservoirs." The Yarmouth Aquarium was altered so as to adjust the storage reservoir
to the show tanks after the plan which exists at
Manchester. Mr. Kent contends that when sea water
is sufficiently clear and aerated it is unnecessary to
spend time and money in the formation of excessive
storage places. Indeed, he goes so far as to observe
that "it is an open question whether, by a trifling
increase of the pumping power, and acceleration of

the circulating stream, reservoirs might not be done away with altogether, substituting in their place a mere well or cistern, for the reception of the water flowing over from the tanks and feeding of the pumps."

The Southport Aquarium has rather a strange combination of the two methods of circulation and aeration which we have been discussing, viz. those at the Crystal Palace and at Brighton; one half being on one plan, and one half on the other. We understand, however, that recently the Brighton method has been given up, notwithstanding the adjacency of the Southport Aquarium to the sea; so that the whole management is now carried on after the Sydenham fashion. The Southport institution has a large number of tanks; and here, as at the Crystal Palace and some other places, the commendatory plan of inscribing the names of the objects on the sides of the tanks is carried out. It is absurd to compel everyone to purchase a guide book before he can understand what fish, &c., are before him! The new arrivals of objects intended for public aquaria are usually placed in special and private tanks, until they are deemed sufficiently acclimatised to be transferred to the show tanks.

In this state of captivity the animals, if healthy, soon feed heartily. For this purpose, at the Crystal Palace Aquarium and elsewhere, twenty-two special tanks are kept in which to preserve alive the crabs, shrimps, fish, &c., intended for food. The introduction

of this live food into the show tanks by the assistants
is always an animated sight, the inhabitants soon
learning the times and appearances of their keepers.
In the Crystal Palace Aquarium, with only 20,000
gallons of water in the tanks, the food supplied to the
living objects costs 120*l.* per annum.

We cannot do better than conclude with the following remarks by Mr. Saville-Kent, contained in his
paper above mentioned as to his experience of the
feeding habits, &c., of the marine animals he has
superintended.   Herrings, whether old or young, are
partial to living food ; feeding chiefly, in the latter
instance, on entomostraca, and the larval young of
the higher crustacea.   Such pabulum being difficult
to obtain so far inland, a variety of substitutes were
offered by way of experiment ; but for a long time
none successfully.   Ultimately an irresistible *bonne-
bouche* suggested itself, in the form of the hard part
or adductor muscle of the common mussel.   This
substance, minced fine, being clean, hard, and white,
with probably a somewhat crustacean flavour, was
devoured with avidity by the little fish, and has
constituted the chief staple of their existence ever
since.   In the course of a few weeks the whitebait
became so accustomed to confinement as readily to
take their prepared food from the keeper's hand—a
circumstance which would seem to indicate that young
fish, like the young of other animals, are more readily
susceptible of domestication, adult herrings not being

known to display an equal amount of confidence towards those who tend them. The food question being settled, another difficulty presented itself, and this time one that threatened, sooner or later, to accomplish the extermination of the whole shoal. Immediately succeeding their advent, a large number of these little fish were found dead each morning, at the bottom of their tanks, a circumstance which at first seemed inexplicable in association with their quiet behaviour during the day. A night inspection, however, happily revealed the cause of their rapid destruction. It was then seen that the nocturnal movements of the herring, at least in confinement, are altogether distinct from those seen by daylight. In the latter instance these movements are very quiet and uniform, the fish swimming round their tank in one shoal and in one continuous stream. At night, on the contrary, the shoal is entirely broken up, each fish taking an independent path, and darting from one side to the other with an amount of agility scarcely to be anticipated by a mere daylight acquaintance with the species. It was during these active nocturnal movements that the fish struck against the rockwork of their tank and came to an untimely end. This mortality, however, was soon arrested by placing a dim light over their tank, which illuminated the outline of the rockwork just sufficiently to enable them to recognise and avoid it. With this dim light the fish still retained their active habits, and it was noticeable that during these night hours they were more than ordinarily alert for food, dashing vigorously at any entomostracan or other minute organism that passed through the water. This circumstance would seem to explain why "drift-net" fishing for herrings can only be carried on successfully at night, that being the time when fish rise to the surface of the water to feed on the innumerable organisms which abound there. They are, in fact, so ardent at such times in pursuit of their food that they needlessly strike into the meshes of the net and get caught, just as the individuals under artificial conditions dash against the rockwork of their tank if sufficient light is not provided for their avoidance. This plan of dimly illuminating the whitebait tank was practised with equal benefit in association with other species that exhibited a tendency to injure themselves during the dark hours of the night, such species again being usually free rangers of the open sea. The picked dog-fish (*Acanthias vulgaris*) is one of these, and is a variety so given to rendering itself an unsightly object by knocking its head against the boundaries of its tank, till it lays its cartilage bare, that it is frequently refused admittance in aquaria. Observations made at the Manchester aquarium, however, revealed that this self mutilation was invariably effected during the night ; and a small light, enabling the fish to see and avoid the rocks, was found an effectual preventive remedy !

Fragments of mussels are usually given to such living marine objects, sea-anemones for instance, as are fixed. The eagerness with which their tentacles close upon them is very remarkable. A pair of wooden

Fig. 116.

Small-spotted Dog-fish (*Scyllium canicula*).

tongs, made after the fashion of sugar-tongs, only with very long arms, are employed to convey the food. But it is astonishing how little food sea-anemones, marine worms, and similar objects require, and we are perfectly convinced there is more danger from over-feeding them than in starving them. When we consider their almost vegetative habits it is evident there can be little loss of tissue from expenditure of muscular force. Expenditure of nervous force there cannot be, for these animals have no nervous systems, or developed in the feeblest degree. The water which continually bathes them contains invisible parts of organic matter, as well as immense numbers of the zoospores of sea-weeds, and is, in fact, what Dr. Carpenter aptly terms it, in the "condition of a very weak broth." Indeed, this "weak broth" is all the food the marine foraminifera, sponges, and many other lowly-organised animals have to feed upon. Sea-anemones, sea-worms, sea-squirts, sea-mats, &c., come in for a share of it, so that the quantity of solid food required to be artificially conveyed to many of them is very small.

CHAPTER XI.

MAMMALIA, REPTILIA, AND FISHES OF PUBLIC MARINE AQUARIA.

THE enormous size of the largest tanks belonging to our public aquaria, and the manner in which the mechanism of aeration and circulation of the water has been perfected, have rendered it possible to exhibit living animals of all kinds whose lives are passed amid aquatic conditions. Hence, such *lung*-breathing animals as porpoises, grampuses, seals, sea-lions, alligators, crocodiles, and turtles, may be now maintained for a time, with almost as much ease as objects of a smaller size. The chief difficulty seems to be, not merely in maintaining these huge creatures in a healthy condition, but in capturing and transferring them uninjured to the tanks. As has already been stated, it is usual to keep the captured animals in places appointed for the purpose, until they are more or less acclimatised, before they are turned out for public exhibition. This transitional stage seems to be necessary in the cases of most animals. Within the next few years our public aquaria will be enriched with many other species of huge fresh-water and marine animals, for one of the tanks at Brighton is almost

capacious enough to admit of the evolutions of a
whale.* A rare species of grampus (*Grampus griseus*)
was placed there in 1875, but unfortunately it only
lived a day. Porpoises are much commoner, and if
there are two or three placed together, they appear to
live under these artificial conditions for a longer time.
Two of these animals lived for five and seven months
respectively, in the Brighton Aquarium, and became
so tame that they would take their food from the
hand of one of the attendants, and came like dogs
from the farther end of their tank at the sound of his
whistle. The habits of animals undoubtedly must
have a great influence on the ease or difficulty with
which they are kept. Some are solitary, others
gregarious, or social. The latter pine or are restless
when alone, but become more cheerful when provided
with companions. Such is the case with the porpoise.
Fish of the herring tribe are its usual food. The seal
is another animal easily tamed, and for years individuals have been conveyed about the country and
exhibited at fairs, &c. It will be remembered that
one species acquired a good deal of notoriety from
its being exhibited in London as the " Talking Fish."
Several specimens have been and still are kept at the
Brighton, Southport, and other aquaria. At Brighton,
however, the chief living objects of interest are the

---

* Owing to the depressed, or horizontally flattened tails of marine
mammals (for the purpose of diving), the tanks are required to be
exceedingly deep if these animals are to be kept healthy.

huge "sea-lions," or eared seals (*Otaria*), from the South Pacific.

The reptilia are not difficult to keep under the artificial conditions of aquarian life, owing to their more sluggish habits. Alligators and crocodiles need darker retreats. At Southport, Manchester, and Brighton, several species of alligators are exhibited, in water cages provided for the purpose. They are fed with a pair of forceps, and take their food without any demurrance to the artificial conditions under which it is offered to them. A living gigantic edible turtle—one of the few *marine* reptiles surviving out of the host of extinct forms which swarmed the seas in the oolitic period—was presented by Her Majesty the Queen to the Brighton Aquarium in 1875, and, we believe, is still kept there. It weighs no less than 3 cwt., and will be easily recognised by the familiar *plastron* or shield on which many spectators have often seen chalked the words, "For soup to-day!" in the London streets. Another species of turtle is that called the "hawk's-bill" (*Chelonia imbricata*). Although in reality a native of the more southerly parts of the Atlantic, this species is occasionally found straying into British waters. It is much smaller than the edible turtle, with the plates of its shield overlapping, hence its specific name. It is these plates that, when polished, go by the name of "tortoise-shell." As regards its diet, it is almost omnivorous, as its strong hawk-like beak indicates, when compared with the weaker mandibles

of the sea-weed-loving turtles. It is a pretty, active
little animal, readily acclimatised, and therefore a
great favourite in public aquaria.

Fig. 117.

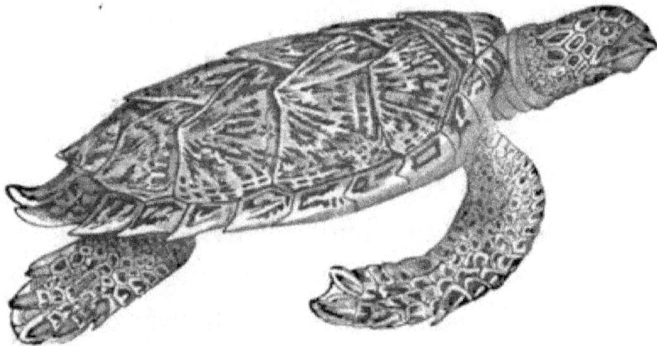

Hawk's-bill Turtle (*Chelonia imbricata*).

Mr. Lloyd's restriction of aquaria to animals that
possess *gills* and not *lungs* is not altogether recognised by nature, unless he refers solely to their
mode of breathing, to which, of course, all the
mechanical aeration of the water is adapted. Reptiles
and mammals of all kinds, are inhabitants of the
same seas as the fishes; and there is no reason why
we should not acknowledge this fact. But, undoubtedly, fishes are more pleasing and familiar
objects in the water than anything else, and as they
are distributed everywhere with an abundance that is
unfailing in its supply, they will always continue to
be among the chief attractions of a marine aquaria.
Few of them are adapted to live in *shallow* or tidal

tanks, which are therefore usually restricted to zoophytes, certain crustaceans, &c. The fishes which
seem to live in shallow water most readily are
the ballan and other wrasses, the rock goby, the
fifteen-spined stickleback, and several others. One
reason why fish will always be favourite objects
is the marvellous variety in their sizes, shapes, and
habits, in addition to their more intrinsic importance
as food. Many people never see anything but dried
fish, or stale fish exposed on stalls ; and these are as
different from the graceful objects seen moving about
in capacious tanks as the stuffed and labelled birds of
a museum are from the winged and animate choristers
of the woods !

The thin, worm-like lancelet (*Amphioxus lanceolatus*)
is kept alive at Sydenham, six of them coming from
Naples alive, in a post-letter! This little fish is not
more than a couple of inches in length, and is a
native of the Mediterranean. So singular is its internal structure (possessing neither brain nor vertebræ),
that an *order* had to be prepared by naturalists for its
separate reception. To zoologists it represents the
embryonic condition of fishes and mammals, in per-
manently possessing a *notochord*. The mud-fish (*Pro-
topterus annectens*) is shown at Brighton, and this object
is even more scientifically interesting than the lancelet,
as being a veritable "missing link" between reptiles
and fishes. It belongs to an ancient class, of which
the most remarkable living forms are now to be found

in some of the Australian rivers. In the early liassic
period these Australian fishes lived in British estuaries,
as is proved by the occurrence of their teeth in the
Rhætic beds of Gloucestershire and elsewhere. The
Brighton mud-fish is nearly related to them, although
a native of African streams. It possesses both rudimentary lungs and gills, whilst its elementary limbs
can hardly be distinguished as legs or fins.

The most interesting of aquarium fishes, however,
are undoubtedly those belonging to the shark family.
Few of them are eaten as food, but many of them are
more or less familiar to the reading public. No fishes
have more graceful motions in the water, on account
of the ease with which their unequally lobed tails
enable them to turn over and round about. They are
very active, especially at night, when they are liable to
hurt themselves against the rockwork, unless some
such ingenious arrangement of dimly lighting the
water be adopted as we have already quoted from Mr.
Saville-Kent. The most familiar of the fishes of this
class which are kept at Brighton, the Crystal Palace,
and elsewhere are the smooth hound, or skate-toothed,
shark (*Mustulus vulgaris*). The tope (*Galeus canis*),
also called the "miller's dog," has been so far tamed
as to bring forth "litters" at Brighton, and the
young (as is usual in some fishes of this class) were
brought forth alive and not as eggs. The largest
individuals of this fish, which is nearly as rapacious,
although not so abundant, as the common dog-fish—

often measure six feet in length. The common dog-fish (*Acanthias vulgaris*) is exceedingly abundant, as every fisherman is aware, and is so easily kept in aquaria that it may be seen in all the public ones.

Fig. 118.

Common Dog-fish (*Acanthias vulgaris*).

Few other fish have such an extensive geographical distribution as this. Fishermen dread it on account of the sharp spines which are placed before the two dorsal, and also behind the ventral fins, for these constitute dreadful weapons of attack or defence. No other species of fish plays such havoc on herring shoals as this, for it mangles what it cannot eat or even kill, and seems to delight in bloodshed and carnage.

Only those acquainted with this fact can understand the bitter hatred which fishermen in general entertain for the dog-fish. When they catch it they will often put it to all kinds of torture, as if in revenge ; and the fish almost seems to know that it has to run the gauntlet of its enemies, and therefore shows no mercy in return, if it can only obtain the chance to injure its foes. It is very abundant off our shores, but we have not heard of its being eaten, except off some parts of the Lancashire coast, and we very much suspect that this is done more from motives of revenge than gustatory enjoyment ! The small-spotted dog-fish (*Scyllium canicula*) is rarer, and of a more graceful shape, so that it is a " stock " object in marine aquaria. Many of these fishes, as well as the skates, have a peculiar way of forming egg-cases, and these are very interesting to the zoologist, because in some instances they are so transparent that he can witness the entire development of the contents through them. These egg-cases may always be picked up on the beach, especially those of the skates, which, on account of their shorter tendrils, seem to be more easily detached from the objects to which the fish originally fastened them. These egg-cases are, however, usually found *empty.* They go by the popular name of "pixy purses." We give figures of those of the common skate (*Batis vulgaris*), and of the smaller spotted dog-fish (*Scyllium canicula*). The latter has very long, pea-like, but hollow tendrils, which wind round the

stems of sea-weeds, stones, &c., the parent fish de-
voting much time and pains to properly fastening
them. The sea water gains admittance to the embryo

Fig. 119.

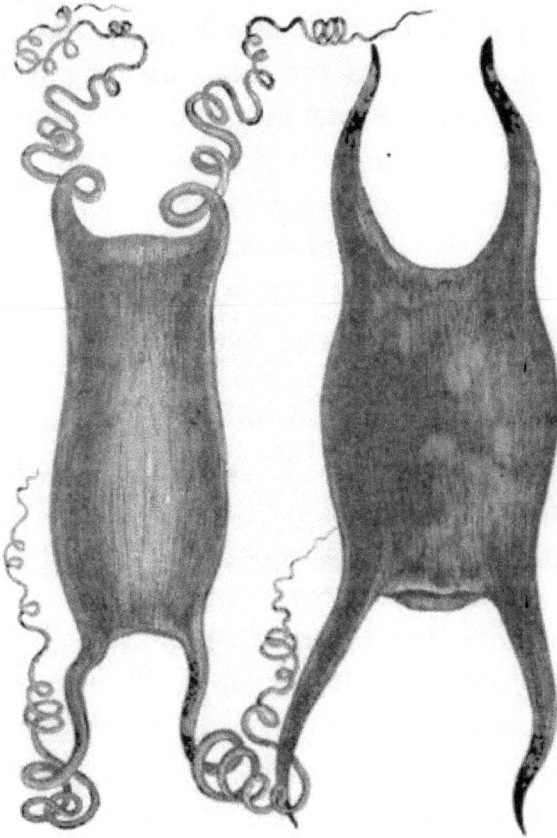

Pixy Purses.

down these hollow tubes. The larger spotted dog-fish
or "nurse hound" (*Scyllium stellare*) are other good
objects for a large marine aquarium. Like the small

spotted, they are natives of deep water, and therefore
require a deep tank. Perhaps this is on account of
their nocturnal habits, as they hide away from the
light in the deep parts where it cannot penetrate,
and are all more or less surface swimmers by night.
In aquaria they are fond of taking shelter in such
nooks and crannies as the rockwork may have provided for them.   Sometimes they are seen lazily
reclining on the shingly bottom, with their bird-like
eye-membranes closed, but the practical announcement
of "meal time" is sufficient to rouse them into a state of
activity only excelled by a batch of hungry carnivorous
animals in a menagerie, to whose lithe and supple
motions, those of the "ground sharks" or dog-fishes
might be compared.

The fact that the latter fishes bring forth their
young in egg-cases, and not alive, as the sharks do,
and the resemblance of these egg-cases to those of the
rays, shows us that the two groups are nearly related.
Perhaps one of the "connecting links" is the angel,
or angel fish (*Rhina squatina*); although why the
popular name should be so easily convertible is not
apparent. There is undoubtedly a strong resemblance
between the rounded head and broad-spread pectoral
fins of this fish and the hooded cowl of a monk. This
fish is tolerably common in the Irish Sea, where it is
shunned by the fishermen on account of its habit of
simulating death, and snapping at them when least
expected. Like many others of its class it is noctur-

nal in its habits, feeding on small flat-fish, crustacea, whelks, &c., indeed almost anything it can obtain. It usually lies half-buried in the sand, which it strews over its body by means of its tail. The females, however, bring forth their young alive; so far, therefore,

the connecting link between the dog-fish and rays is destroyed. The shape of these fishes, nevertheless, is a good indication of their affinities to the skates or rays. At the Brighton Aquarium the angel fishes have always done remarkably well. The thornback or skate (*Raia clavata*) is one of the best known of our flat-fishes, and is much eaten by the poorer classes of this country. Those who have partaken of the flesh of the *wings*,

Fig. 120.

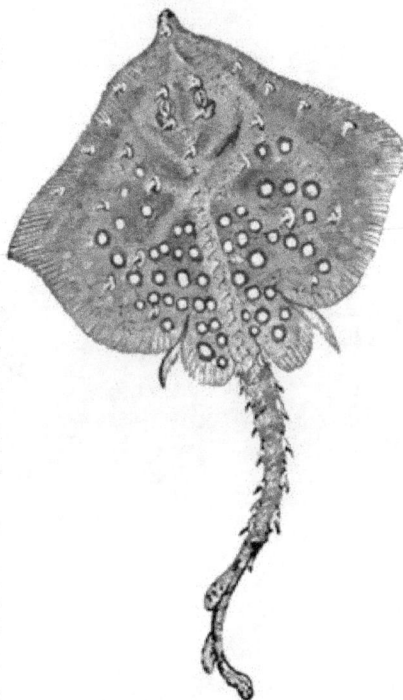

Thornback (*Raia clavata*).

or broad pectoral fins, will, however, be inclined to say that it is "a dish for a king." The thornback is easily acclimatised, and will live in the same tank with cod, dog-fish, sturgeon, &c. The homelyn or spotted ray (*Raia maculata*) is also eaten as food.

This is usually a smaller, but much more graceful fish than the former. In it the spines are not distributed over the body, but are confined to a single

Fig. 121.

The White Ray (*Raia lintoa*).

series running down the back and tail. The common skate (*Raia batis*) is a different species from either, and its flesh is usually preferred to theirs. On the eastern coasts some very large individuals occur, and

indeed it is the stock fish of that district. The
Brighton Aquarium contains all the foregoing species,
as well as the white or sharp-nosed ray (*Raia lintoa*).

The cat-fish (*Annarhicas lupus*) is another singular
species about which fishermen have "spun yarns" for
untold years. It is sometimes called the "wolf-fish,"

Fig. 122.

Cat-fish (*Annarhicas lupus*).

and this is acknowledged in its specific name. Its
savage appearance, caused by the fierce display of
naked teeth, do not belie its habits; for when it
is caught it will snap at and seize any object the
fishermen may present to it. In addition to these
teeth the cat-fish also possesses roundish palatal
teeth, set close together like a miniature pavement of

boulder stones. These are for the purpose of crushing the shells of the crabs and other crustaceans on which it most delights to feed. It belongs to the usually harmless family of the blennies, one of which, the little smooth blenny (*Blennius pholis*), derives its name from a Greek word signifying "slime," on account of the abundant mucus which, in common

Fig. 123.

The Blenny (*Blennius pholis*).

with all fishes, it secretes from the medial line of scales, which are perforated to allow the mucus to exude, and thus decrease the friction caused by rapid movements in the water. Many larger fishes support parasites on their scales, and these are usually crustaceans which have undergone a very strange degradation. Of course, they are popularly spoken of as "fish-lice," and have been regarded as unhealthy signs on the part of the fish. Any fishmonger, however, will tell you that the fish thus affected are usually stronger and better than those without them, and Van Beneden has shown that the parasites really live on the excessive mucus, which they thus check and prevent decomposing; so that they are actually serviceable to fishes. The smooth blenny is a good shallow tank object, and may be kept and even tamed there easily, especially if the tank represents the hiding-places of the rock-pools in which it delights when free. The male, like the stickleback, will defend the eggs against enemies.

Some of the blennies are really very attractive objects, especially the species called the "butterfly blenny" (*Blennius ocellatus*), which is kept at the Crystal Palace, Brighton, &c. Its dorsal fin is largely de-

Fig. 124.

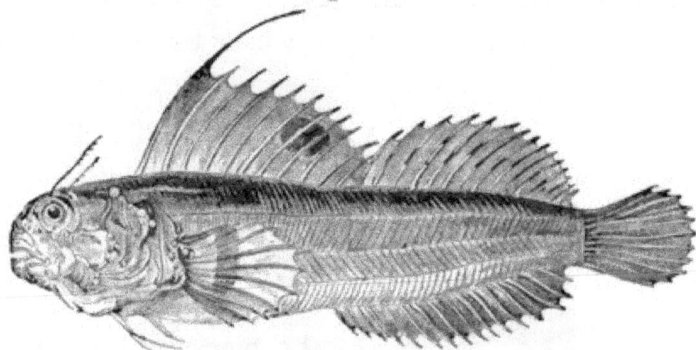

Butterfly Blenny (*Blennius ocellatus*).

veloped, and is seen to great advantage when the fish is swimming about, the dark blue or brown spots with which it is adorned giving it something of a butterfly appearance. It is not by any means so common in our seas as the viviparous blenny (*Zoarces viviparus*), a fish which goes by the name of the "green-bone" along the eastern coasts, where it is very common, on account of the deep-green colour of the bones when the fish is cooked. This species departs from the usual habit of fishes in bringing forth its young alive, the reason being that the ova are kept longer in the body of the female. The young are very pretty objects, especially under the microscope, where the circulation of the blood is plainly visible. Small though they are, they seek their own food as soon as they are born. Some of the blennies can wriggle out of the water by means of the free rays of the ventral fins; and the smooth blenny is especially fond of climbing on the rockwork of the tanks just above the water level. So tamable is this species that Mr. Lloyd says it will even feed when taken into the hand.

The gobies are a useful and interesting series of aquarium objects, on account of the ease with which they can be kept in a state of healthy activity. The largest species is the black or rock goby (*Gobius niger*), which may be seen in every public aquarium in a very tamed condition. It has a kind of "sucking fin" on the ventral or under surface, by means of which it can attach itself to the smooth face of the glass, on which it sometimes deposits vast quantities of minute eggs, and then watches them most pertinaciously. It abounds in our rock-pools, and may be easily captured there by means of a hand-net.

Fig. 125.

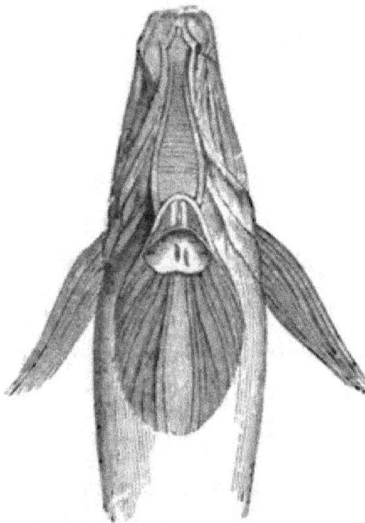

Ventral surface of the Rock Goby (*Gobius niger*), showing sucking fin.

Another species, the spotted goby (*G. minutus*) is much smaller in size, and very closely resembles the sandy bottom in which it hides, so that it often requires a keen eye to detect it when hiding up. Among fishermen this species goes by the name of " polewig," or " pollybait," and as it abounds in the lower part of the Thames, it goes largely to make up the somewhat heterogeneous piscine compound known as " whitebait " !

CHAPTER XII.

FISHES FOR MARINE AQUARIA.

THE most attractive marine fishes kept in our public aquaria are unquestionably the wrasses, gurnards, and dragonets. Many of them are resplendent with the most beautiful colours and tints, which rival the hues of tropical birds and insects ; and as they all bear confinement well and are not difficult to procure, we cannot be surprised at their being aquarium favourites.

Fig. 126.

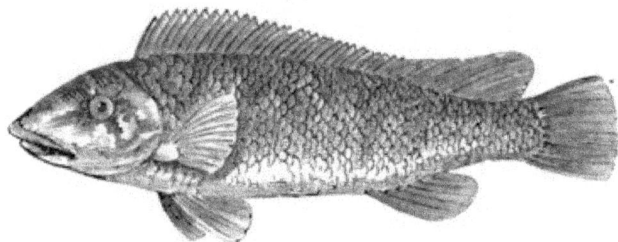

Ballan Wrasse (*Labrus maculatus*).

The wrasses are crustacean and shell-fish feeders, as a rule, and their teeth are peculiarly adapted for picking shells off rocks. The ballan wrasse (*Labrus maculatus*) usually assumes mingled colours of blue and green ; occasionally it will be adorned with russet-brown, orange, and yellow. It is this chameleon-like power

of assuming fresh colours that has undoubtedly multi-
plied the species beyond requirement, and hence we
find more synonyms among the wrasses than in any
other group of fishes. In addition to the prominent
rows of front teeth, the wrasses have the power of
elongating the jaws, and also possess protrusile lips,
whence their generic name of " labrum," a lip. These
peculiarities as well as their tints have caused them
to be termed " sea-parrots" among fishermen—another
name they bear is that of the " old wife." The ballan
wrasse is the commonest species, and attains the
largest size, often weighing three or four pounds.
Frank Buckland tells us, in his charming work ' Popular
History of British Fishes,' that this species on the
north coast of France is usually red, and there goes
by the name of the " red old woman." The adjective
name of colour changes with that of the fish to green
and yellow, but the "old woman" remains. From
what Mr. Couch says, these fish appear to have habits
not unlike those of the common stickleback, the
largest individuals being lords of their respective
districts. The red wrasse (*Labrus mixtus*) female, is
distinguished by the three dark and four light rose-
coloured spots which appear at the base of the hinder
part of the back fin. These vary in number, so that
this species is also called the "double spotted," when
there are but two dark spots. The rest of the body
is usually a fine red on the upper, and a pale orange
on the lower part of the body. Mr. Kent tells us that

O

the best food for the aquarium wrasses is the common
shore crab (*Carcinus mænas*), the ballan wrasse being
so fond of it that specimens dropped into their tank

Fig. 127.

Blue and Red Cuckoo Wrasse (*Labrus mixtus*), male and female.

are torn into pieces before reaching the bottom ; but
Mr. Lloyd states that shrimps are their most favourite
food.   The colours and markings of the male and
female wrasses generally differ from each other, and it

has been found from aquarium experience that of the
two species named the "red" and "blue," the former is
the *young* fish.* This colour is also that of the female,
so that the fact of highly-adorned male. fishes assuming the tint and colour of the female when young,
is analogous to the rule which prevails among many
birds, of the young male bearing the plumage of the
adult female. The cuckoo wrasse (*Labrus mixtus*) is
to be seen in public aquaria, the males and females
in the same tank, being both highly coloured, but
so different in tint and marking that until recently
they were regarded as different species.

The corkwing and rainbow wrasses are distinguished from those just named by the edge of the
gill-cover being finely toothed or crenated, hence

Fig. 128.

Corkwing Wrasse (*Crenilabrus melops*).

their generic name of *Crenilabrus*. The true wrasses
are thus marked when young, but lose the teeth as
they reach their adult condition. The corkwing (*Crenilabrus melops*) is not a very common British fish,

* *Labrus variegatus* and *Labrus trimaculatus* are the male and female
of one species, *Labrus mixtus*. The green wrasse (*L. lineatus*) is the
young of *L. maculatus*.

smaller than the species just mentioned, but not less beautiful in its bright green body, variegated with a net-work of scarlet and blue. Its fins are of the brown colour of cork, whence the common name of this fish. The rainbow wrasse (Fig. 129) is a very rare fish, whose name is borrowed from its prismatic or rainbow tints. As the specific name implies, it is very abundant in the Mediterranean.

Fig. 129.

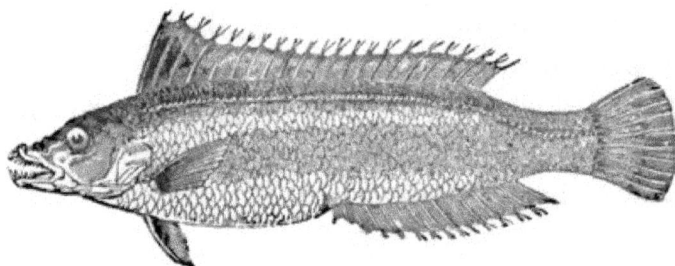

Rainbow Wrasse (*Julis Mediterranea*).

The dragonet, or yellow skulpin (*Callionymus lyra*) is another common aquarium fish, whose gorgeous colours and dragon-like fins and appearance are sufficient recommendation for its introduction. It is not a common British fish ; but this may be because it is fond of deep. water, and of keeping at the bottom. Like the wrasses, the male and female are differently coloured, and were formerly believed to be different species. The male is easily distinguished by the very long dorsal fin ray. Their green eyes gleam with a beautiful fire-like expression, and are pro-

minently fixed on the head. The head is usually
striped with blue, on a yellow ground. The dusky
dragonet is the female, and was formerly called
Callionymus dracunculus. It is a much commoner,
and less beautiful, British fish, having, as its popular

Fig. 130.

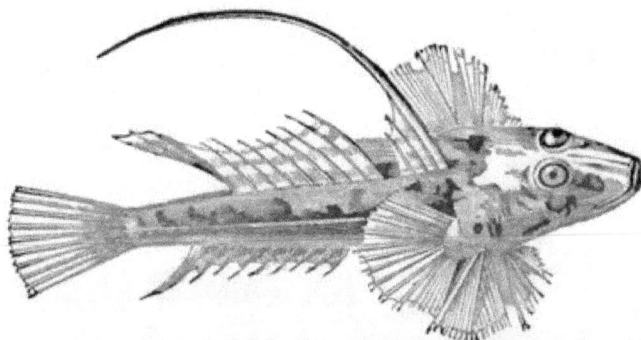

Yellow Skulpin, or Gemmeous Dragonet (*Callionymus lyra*), male.

name imports, a dingy hue. The long dorsal fin is
absent in the female. Its colour is usually a reddish-
brown, and therefore it goes by the name of the
" fox."

The " angler fish" (*Lophius piscatorius*) is a creature
possessing such a singular structure and habits that
we cannot wonder it is a favourite object in our public
aquaria. Sailors and fishermen are never at a loss in
drawing comparisons, and every animal which has
attracted their attention usually rejoices in several
names. This is the case with the angler fish, also
called " toad fish," " sea-devil," and " fishing frog."
In shape it is not unlike a huge tadpole, and one of

its peculiar organs illustrates to us the case with which nature sometimes modifies, transforms, and specially adapts one part of the animal frame to an extraordinary use.   The back or dorsal fin in fishes

Fig. 131.

Angler Fish (*Lophius piscatorius*).

is usually supported by fin rays.   Not unfrequently one or more of these is larger than usual (as in the dragonet and weever), and then becomes a weapon of offence or defence.   But in the angler fish, the first three spines are modified into long *tentacles*, and removed from the back to the head (see Fig. 131).

Bending gracefully over in front of the mouth, the animal now employs these modified spines as artificial baits. The fish is concealed in the mud, but gently waves these organs about as if they were moving sea-worms, in order to attract smaller fishes on the look-out for food. The front ray is slightly clubbed at the end, and adds to the tempting look of the living bait. The mouth of the angler fish is extraordinarily large, and is armed with a row of teeth that would certainly deter even the most foolish of little fishes, and in spite of the most attractive of baits, if they beheld it. As we have seen, however, owing to its habit of burying in the sand or hiding among the rocks, the head is concealed. Mollusca as well as crustacea are equally fair food to the angler fish, whose palate is set all over with pavement-like teeth, for the purpose of crushing the shells and carapaces.

The lump fish, or " lump sucker" as it is also called (*Cyclopterus lumpus*), is further known by the names " sea-owl," " cock-paddle," " sea-hen," &c. It is a very abundant fish on the southern and eastern coasts, and its bright colours and unfish-like form make it a favourite object in the aquarium. Few people would recognise it as a British fish when alive, for, although we have frequently seen it exposed on fishmongers' stalls, the colours soon fade after death, and the semi-translucent look it has when alive in the aquarium is utterly gone. Its popular name is derived from the saucer-shaped disk, which may be seen

between the two pectoral fins. This is a sucking
disk, and by its means the *Cyclopterus* can bring itself
to an anchor in a strong current, where perhaps it

Fig. 132.

Lump Sucker (*Cyclopterus lumpus*).

might fall a victim from its own weak swimming
powers. Its other name of " sea-hen " is also well
deserved, for its young—which are most lovely little
creatures—sportively follow the mother-fish about as

if they were chickens. Their frolicsome ways make
the young exceedingly interesting and amusing
objects. The male fish is believed to " mount guard"
over the eggs, after the fashion set by the little
stickleback. Some of the lump fish often weigh ten
or twelve pounds when full grown.

Speaking of the stickleback reminds us of a
common species, generally known as the "fifteen-
spined stickleback" (*Gasterosteus spinachia*), the largest

Fig. 133.

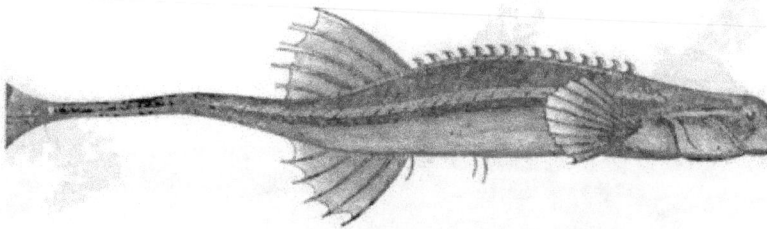

Fifteen-spined Stickleback (*Gasterosteus spinachia*).

of its kind. This pretty fish is never more than six
inches in length, but its body is exceedingly long in
comparison with its breadth, so that it can dart
through the water with the rapidity of an arrow.
Like its fresh-water relative, this fish constructs, and
even stitches and glues together, a nest, usually of
sea-weed, and the male defends the eggs laid there so
determinedly that he appears to be an animated
arrow, constructed for the special purpose of "charging"
other and huger fishes desirous of making a meal of
the nutritious ova. Its spines are capable of inflicting

a severe wound, both on fish and man, whence its name among the west-country people of " sea-adder." It is the two rows of elongated hard plates underneath the bodies of sticklebacks which have given them the name of *Gasterosteus*, i. e. " bony-belly." The fifteen-spined species is a shore-loving fish, and should therefore be kept in a shallow or tidal tank. The " pogge," or armed bull-head (*Aspidophorus Europæus*) is not distinctly related to it, but its body is

Fig. 134.

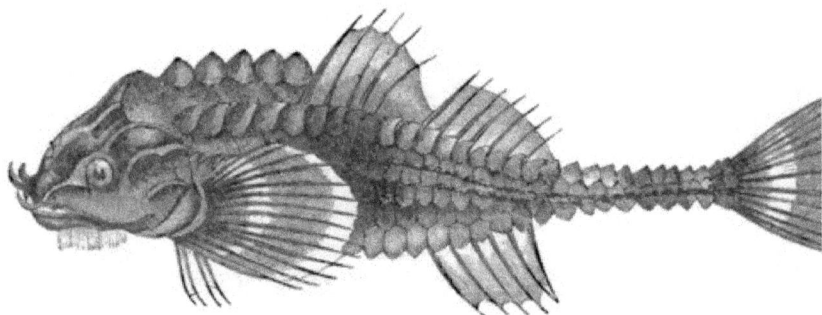

Armed Bull-head, or Pogge (*Aspidophorus Europæus*).

more octagonal, being covered with eight rows of strong plates. The mouth is furnished with curious cilia. This fish is a very graceful species, but terribly destructive to the young shrimps, prawns, and lobsters. In pursuit of the latter species they frequent the deeper parts of the sea, and in the aquarium may be seen keeping close to the floor, whose tint so well comports with that of their bodies that it screens them from observation. Indeed, to a certain extent the pogge has the chameleon-like power of changing its tints according to the ground it haunts. Few fishes are therefore better defended against enemies, if we take into account the mail-clad body and the deft means of concealment they have by simulating the colour or tint of their hiding place.

In many respects the gurnards are aquarium favourites. The elegant shapes and usually bright colours and tints of our British species would be quite sufficient to induce the naturalist to place them in his "show tanks." Those who have seen the larger sapphirine gurnard alive will not soon forget the exceedingly bright colours on the very large pectoral fins, which have earned for it its popular name. Some of the species are called "Butterfly Gurnards" from their coloured spots, streaks, and tints. Their thin and tapering bodies contrast strongly with their large and somewhat angular heads. All of them have well-developed pectoral fins, and it is these, when coloured, that cause them to have something of a butterfly appearance whilst swimming. We have seen how in the angler fish three of the dorsal fin rays are modified both as to their use and position, so that they serve as natural angling baits by which the fish attracts its prey. In the gurnards we have a modification of the first three of the rays of the pectoral fins, which are actually used as legs, the gurnards being able to creep along the sea-floor for some distance by their aid. Their

mode of walking very much resembles that of the
lobster upon its slender jointed feet. These improvised feet also disturb many kinds of animals hiding
under the sand, which soon fall a ready prey. The
streaked gurnard (*Trigla lineata*) possesses exceedingly rich tints, especially when young, insomuch

Fig. 135.

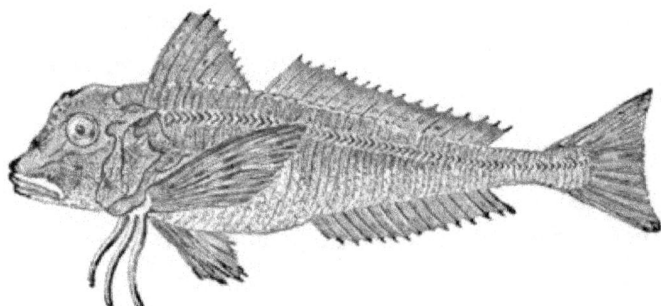

Streaked Gurnard (*Trigla lineata*).

that in the water it suggests the appearance of some
gaily-coloured tropical bird. Its usual adult length
is about 12 inches. The head, mouth, back, dorsal
fin, and tail are of a vermilion colour ; and sometimes
of a dusky red. Its eyes are of a most lovely and
brilliant blue. The dorsal fin is marked with bars of
red and pinky clouds. The medial line is sometimes
a deeper red, whilst the large red pectoral fins are
further adorned with spots and markings of green and
blue. The spots are most numerous in the older fish,
and fewer and larger in the younger. The grey
gurnard (*Trigla gurnardus*) is a very common fish,

occurring on our coasts in extensive shoals, which are much sought after by fishermen, on account of the demand for this kind of food among the lower classes.

Fig. 136.

Grey Gurnard (*Trigla gurnardus*).

Its pectoral fins are not so large as those of the streaked gurnard, and its colours are less brilliant and striking. The piper (*Trigla lyra*) has a comparatively larger head than the preceding species, whilst its body

is more slender and tapering. It is very abundant
on the southern and western coasts of England.

A fish which bears a bad reputation among the
fishermen is the lesser weever, or "sting fish"
(*Trachinus vipera*). It is said to lie on the bottom of

Fig. 137.

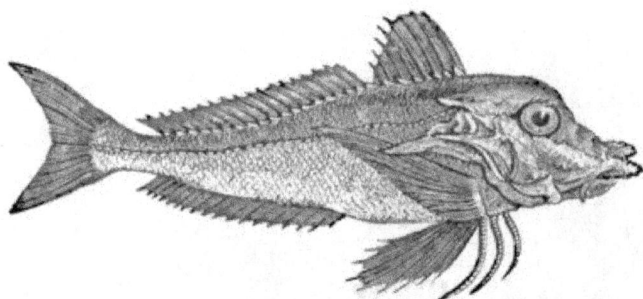

Piper (*Trigla lyra*).

Fig. 138.

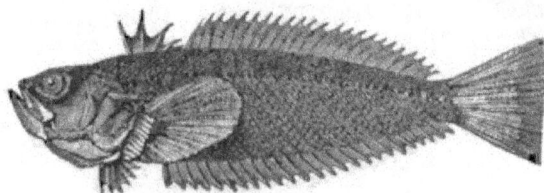

Lesser Weever (*Trachinus vipera*).

the boat quiet enough until something or somebody
comes within its reach, and then to bound up and
drive the sharp thorn-like rays of its dorsal fin into
its enemy. Dr. Gunther has proved that the fishermen are right in holding that the wound thus made
in the flesh is poisoned, and that the double-grooved

dorsal spines are poison organs, the poison being contained in cavities of the spines until poured out.* It is
not a very beautiful fish, its head being snakish-looking
and ugly. The body is adorned with slanting greenish-yellow bars, and is seldom more than 6 inches in
length ; so that it is not to be confounded with another
species, the greater weever (*Trachinus draco*), sometimes called the "dragon fish." The latter bears sharp
strong spines on the upper part of its gill-covers, but
they have not been scientifically discovered to be
poisonous, although popular belief declares they are.
Bathers are not unfrequently wounded by this fish's
spines. The former species only is usually kept in
aquaria, but at the Crystal Palace the *draco* is also
exhibited. Both species are fond of burrowing in the
sandy floor of the tank.

The basse (*Labrax lupus*) is a sea-perch, and
usually thrives well in aquaria. Its body is very
symmetrically shaped, and it seems to take pleasure
in keeping its large silvery scales clean by rubbing
itself among the sand and shingle. It is sometimes
sold as "white salmon," for it much resembles that
fish in shape and colour, but it is not spotted. The
scales, also, are much larger, and the flesh is *white*,
and in taste resembles that of the turbot. The
Romans seem to have been much fonder of this fish
than we are, and there can be no doubt whatever as

---

* Mr. Lloyd's experience of this fish, in aquaria, does not bear out the
character given to it by our fishermen.

to their superior judgment concerning it as a food-fish. The specific name of *lupus* (a wolf) was given to the basse on account of its hunting its crustacean food in packs. It is a good fish for summer sea-angling, either from open boats or the ends of long piers.

CHAPTER XIII.

FISHES FOR THE MARINE AQUARIUM.

THE so-called "migratory" fishes, as the mackerel, herring, pilchard, sprat, &c., are more difficult to keep long in active health in our marine aquaria than those which prefer to pass their lives always in the same locality. All of them are gracefully-shaped fish, and the mackerel, in addition, is one of the most beautifully marked and coloured. Their well-known value as food-fishes, and the fact that everybody is familiar with their appearance, was a strong inducement for the managers of the first-formed aquaria to exhibit them in their show tanks. All of them are more active by night than by day, and we have seen how Mr. Henry Lee and Mr. Saville-Kent ingeniously prevented the young herrings from mutilating themselves. When mackerel were first introduced into the Brighton Aquarium in 1872, some of them beat themselves to death by dashing against the rockwork. Since then they have been acclimatised, and living specimens may now be seen there which have lived ever since, and grown considerably in size. Herrings are exceedingly active and graceful fish, and were first domesticated at the Brighton Aquarium, and

P

specimens are there to be seen which have been living in confinement nearly three years. The sprat is a different species, although formerly regarded as the young of the herring. It is even more difficult to keep under artificial conditions than the latter fish, although it has been successfully effected at Brighton, Manchester, and elsewhere. The pilchard is annually taken in immense quantities off the Cornish coast, but few of them are consumed in England, Italy being the market for them. In its young state it is known as the "sardine," and in that condition is imported into this country, preserved in oil, in the well-known tins. The specimens in the Brighton Aquarium were caught off the Sussex coast, for sometimes the pilchard strays along the south-eastern and eastern coasts. Just now it is being discussed whether adult pilchards preserved in oil, like sardines, could not be turned to home use, so as to give us an additional food-fish, and one that could not fail to be appreciated. The adult fish is said to be quite as good when cured and preserved in this way, as the young, or sardine.

The whiting and cod are both good aquarium objects, and are almost as familiar to the general public as herring and mackerel. We may find them in all public aquaria. The former (*Gadus merlangus*), like the cod, is gregarious in its habits, and is said to be quickly accustomed to confinement. It is exceedingly pretty to see them gracefully making the circuit of their artificial home, their silvery flanks catching and reflecting the subdued light like mirrors. The pollack whiting (*Gadus pollachius*) is a nearly-allied species, exhibited at the Crystal Palace, Southport, and Brighton. This species is more solitary in

Fig. 139.

Pollack Whiting (*Gadus pollachius*).

its habits, and more beautifully coloured with rich yellow. The whiting pout (*Gadus luscus*) is also kept at the Crystal Palace, Southport, and Brighton. Those at the latter place are remarkably tame, and will take their food from the attendant's hands. The natural habitat of this fish is among the rocks, whence its other name of " rock whiting."

The codfish (*Gadus morrhua*) is nearly related to the foregoing, but attains a much greater size, some having been taken weighing sixty pounds. As is well known, it is an exceedingly prolific fish, Mr. Frank Buckland giving as an example one specimen in which the removed roe weighed seven pounds and three-quarters, and was calculated to include nearly seven millions of eggs. From aquarium experience it has been discovered that Professor Sars was right in his surmise that the spawn of the cod was not deposited along the bed of the sea, as was formerly believed, but floats on the surface of the water during the entire period of its development, which occupies

Fig. 140.

Codfish (*Gadus morrhua*).

about sixteen days. In the Crystal Palace and Brighton Aquaria this spawning has repeatedly taken place, and many important investigations have there been made in the embryology of the young fish. On account of its great prolificness, it is one of our commonest and best sources of fish-food, and it is

highly important we should do all we can to keep
it such. Its chief food is marine worms, small crustacea, and small mollusca ; and the carapaces and
shells of the two latter may be found in its stomach
in every stage of decomposition and dissolution. The
cod becomes very tame in confinement, and is said
even to manifest signs of attachment to those who
feed it. The three-bearded rockling (*Motella tricir-*
rata), as well as the four and five-bearded rocklings,

Fig. 141.

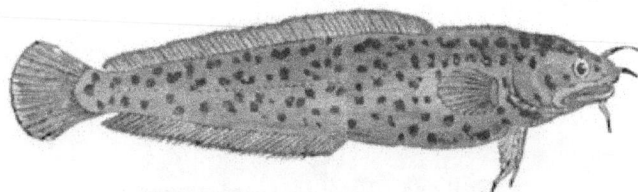

Three-bearded Rockling (*Motella tricirrata*).

are not distantly related to the cod and whitings.
They take their popular names from the barbels
which hang from the lower jaws, and which are possessed by the cod family generally. All of them are
nocturnal in their habits, and one of them, the five-
bearded rockling (*Motella mustela*) builds a nest for
her eggs in the crevices of the rockwork, made up
of corallines, sea-weeds, &c. The haddock (*Gadus*
ægelfinus) is another well-known species of cod, now
to be seen in some public aquaria. The coal-fish
(*Merlangus carbonarius*) is so named from the black
colour it frequently assumes. It is abundant in the

Baltic and the northern seas, and to the inhabitants
of the Orkney Islands its young are the chief food.
Among the Irish and Scotch fishermen it goes by a
variety of names. It is a great enemy to salmon
smelts, as well as to young herrings, as many as

Fig. 142.

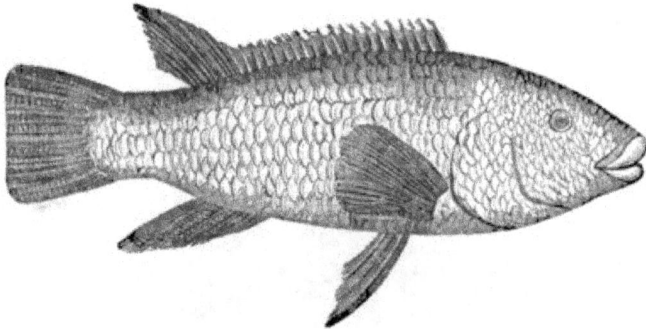

Coal-fish (*Merlangus carbonarius*)

Fig. 143.

Grey Mullet (*Mugil capito*).

twenty-six salmon fry having been taken from the
stomach of a single coal-fish.

The grey mullet (*Mugil capito*) is another familiar
food-fish, domesticated in most aquaria. Not unfrequently it will leave the sea water for a short

period, and ascend a river, like a salmon. Possibly
this irregular act on the part of the grey mullet and
some other fish may be to get rid of superfluous
parasites, for the fish always return apparently better
for the change. This fish has been a deserved
favourite at the table since the time of the Romans.
Its food mainly consists of dead and decaying vegetable matter, eked out occasionally by worms, small
crustaceans, mollusca, &c. It is very fond of browsing
on the green confervæ in the tank, hence it is a valuable vegetable scavenger in aquaria. Not unfre-

Fig. 144.

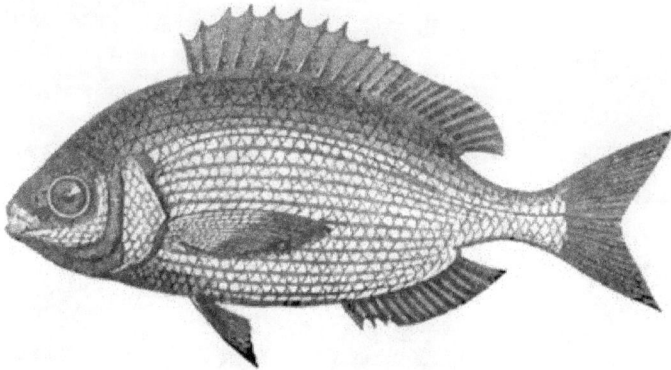

Black Sea-bream (*Cantharus griseus*).

quently it attains a weight of fifteen pounds, although
its average length is about eighteen inches. The black
sea-bream (*Cantharus griseus*) is another handsome
aquarium fish, possessing a bright silvery hue, and
remarkable for its delicate spreading pectoral fins.

Its popular name is derived from the singular change
which affects the male at the breeding season, when
the silvery scales are overspread by a sooty blackness.

As a matter of course, that magnificent fish the
sturgeon (*Acipenser sturio*) is to be seen alive at the

Fig. 145.

Sturgeon (*Acipenser sturio*).

Manchester, Brighton, and most public aquaria of
importance.   One at Manchester was nearly 9 feet
in length and 4 feet in girth, and was the biggest
introduced.   It became excited, however, at the
smallness of the space allowed it, in comparison to

the freedom of its natural habitat, and so died soon
after its introduction. Its successor was somewhat
smaller, but is alive at the time we are writing.
Much the same luck occurred to the first Brighton
sturgeons. This fish belongs to that ancient family
the *Ganoids*, although, unlike one division of them,
it is not completely clad in bony armour, but has
four rows running along the body, and one large row
along the medial line. Its name of "royal" is derived
from an unrepealed Act of Edward II., whereby this
fish become the property of the sovereigns of England.
Like the salmon, it can live in fresh as well as salt
water, and, in the north, where it attains an enormous
size, it is usually found in the large estuarine rivers.
The well-known substance called "caviare" is prepared from the roe of the sturgeon. The sterlet
(*Acipenser rutheorus*) is also exhibited at the Manchester and Brighton Aquaria. This fish was formerly
believed to be the young of the former, but it is now
known to be a distinct species. Its five rows of bony
plates occupy much the same position, and it also
resembles the royal sturgeon in its general habits.
This species is very common in the river Volga.

The pipe-fishes (*Syngnathidæ*) have long been
favourites in aquaria, and some of the species are
kept in all our public institutions. Like their rela-
tives the sea-horses (*Hippocampus*), the entire surface of their bodies is covered with angular bony
plates, so that they can be kept without stuffing

after death, and not lose their shape.   The largest
is the great pipe-fish (*Syngnathus acus*), which never
exceeds 18 inches in length.   All the pipe-fishes
swim in a nearly vertical position, with their bodies
very rigidly disposed, and locomotion is effected almost entirely by the very rapid undulation of the
pretty dorsal fin.   Their colour is usually palish brown,

Fig. 146.

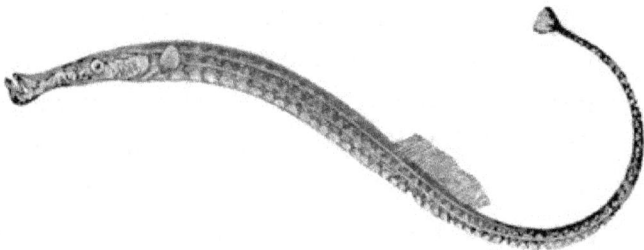

Deep-nosed Pipe fish-(*Syngnathus typhle*).

but they are marked by dark brown bands.   All of
them love to hide among the beds of " sea-grass "
(*Zostera marina*, a true flowering plant, and not a
sea-weed), although their food consists of the smaller
crustacea, for they cannot take much of any other on
account of the peculiar structure of their mouths,
though they sometimes eat their own young.   The
diameter of the body of the great pipe-fish rarely
exceeds half or three-quarters of an inch. The slender
upper and lower jaws are united, and open only just
in front.   The different species of pipe-fish are easily
recognised by the variations in the head and tail.   In

some the latter has no caudal fin, and the tail is then prehensile, like that of the *Hippocampi*, or "sea-horses." In others it expands into a beautiful fan-like object.

As these fishes move about solely by means of the dorsal fin, and not by inflections of the body, the

Fig. 147.

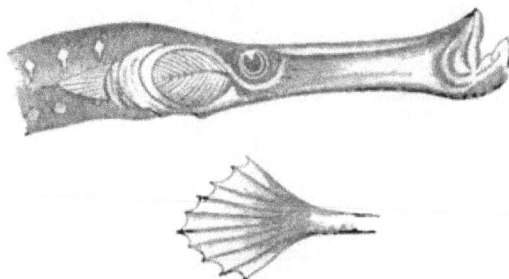

Head and Tail of Broad-nosed Pipe-fish (*Syngnathus typhle*).

Fig. 148.

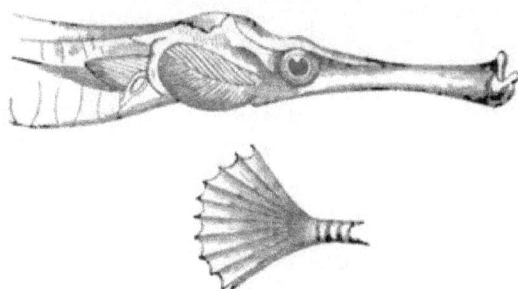

Head and Tail of Great Pipe-fish (*Syngnathus acus*).

reader may form some idea of their graceful and gliding movements in the water. One of them (*Syngnathus æquoreus*) has been seen swimming far out at sea, whence its name of the "Oceanic Pipe-

fish." The most singular feature about these pretty
creatures, however, is that the male fish is provided
with a pouch, formed by an infolding of the skin of
the lower surface of its body. Into this pouch the
female deposits her eggs, which are there fecundated
by the male and carried about by him, kangaroo-
fashion, until they are hatched! It is not true, how-

Fig. 149.

Head and Finless Tail of Oceanic Pipe-fish (*Syngnathus æquoreus*),
nat. size.

Fig. 150.

Head and Finless Tail of *Syngnathus anguineus.*

ever, that the young fish return to this pouch for
shelter, as has been stated ; the analogy to the
marsupial animals being there carried too far. The
one species of sea-horse rarely found in British seas is
usually to be seen in the same tanks as the foregoing,
in company with a Mediterranean species. Their
popular name is derived from the striking resemblance
of the outline of the head and neck to those of a

horse. The two species are *Hippocampus brevirostris* and *H. ramulosus*, the latter much rarer than the former. At first there was experienced a great difficulty in getting proper food for these objects, but it is now partly overcome. Their tails are prehensile, and may be seen twisted round some coralline or other marine object. They seem to possess the means of communicating with one another by means of *sound*, according to the observations of Mr. Saville-Kent at the Manchester Aquarium, where one species has bred.* These communicating sounds are short, snapping noises, produced by a complex muscular contraction and sudden expansion of the lower jaw. When moving about, the sea-horses employ only their transparent, fan-shaped, dorsal fins, which work them along on the principle of the screw-propeller, the movement being quite rythmical, owing to each of the fin rays striking the water in succession. Its favourite food is the minute opossum shrimp (*Mysis chameleon*), and this it

Fig. 151.

Sea-horse
(*Hippocampus
brevirostris*).

will give chase to in a very stately fashion, uncoiling its prehensile tail, and moving towards its prey by means of the propelling action of the dorsal fin, with never-failing dexterity. When within half an inch of

---

* Upwards of a thousand young Hippocampi were recently bred in the Southport Aquarium.

the opossum shrimp it opens its mouth, and inflates
its cheeks so as to cause an in-rush of water, down
which the unfortunate crustacean is engulphed. The
opossum shrimp may be seen swarming in the water

Fig. 152.

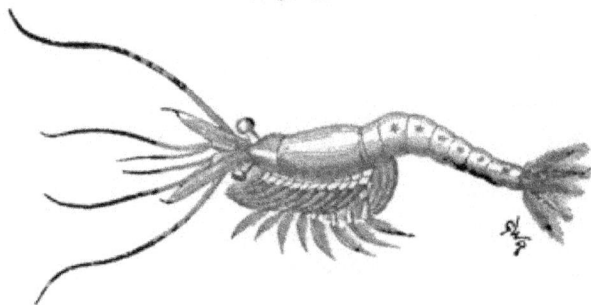

Opossum Shrimp (*Mysis chameleon*).   (Three times natural size.)

Fig. 153.

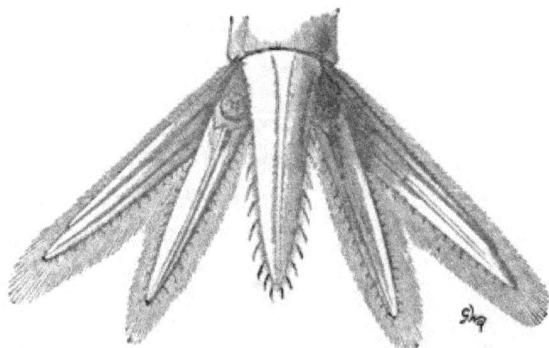

Tail of Opossum Shrimp with *otoconia.   a. otoconia.*
(Magnified ten times.)

of most rock-pools. It is a queer little object, with
the very singular peculiarity of being able to hear by
its tail, at the base of which are certain cavities filled
with crystalline *otoconia*, or "ear-bones."

The well-known " flat-fishes," such as the sole, dab, flounder, plaice, brill, turbot, &c., are in strong force in all our large marine aquaria, where they attract much attention on account of their unexpected graceful movements in the water. It is only there that we can thoroughly understand the marvellous changes which have been effected in the life-history of these fishes, by means of .which both their structure and habits have been slowly adapted to their present condition. Their popular name of " flat-fishes " is apt to lead one astray in his conclusion as to the cause of this flatness. It is not due to the fishes lying with the belly or ventral side downwards ; but to their habitually lying on one *side.* They are *compressed,* not *depressed,* as in the skates. Then only one side is coloured, that next the ground being very light. But the coloured part is usually called the "back," whereas we now see it is only a special adaptation to the really wonderful modifications these animals have experienced. Moreover, the observer may notice how soles and other flat-fishes having a uniform dull neutral tint usually settle down on the *sandy* part of the bottom of the tank, and add still further to their concealment by dusting themselves over with the fine sand, which rises and partly settles down over them, so as to form an admirable screen. Such flat-fishes as the plaice, on the other hand, will be often seen to select that part of the bottom where gravel is abundant ; and then we may notice the meaning of the ochreous or yellow spots distributed over the surface of one side of such fishes. They simulate the bright pebbles, and help to conceal the fishes from their enemies.

In the plaice (*Pleuronectes platessa*), and sole (*Solea vulgaris*), the coloured or upper surface is the right side; whilst in the turbot (*Rhombus maximus*) it is the left. In both cases, whether the upper side be

Fig. 154.

The Flounder (*Pleuronectes flesus*).

right or left, the eyes are turned round so as to be placed on the upper side. When swimming in the water the same abnormal position is adopted, the fishes swimming on their sides, and not erect, after the usual manner. The apparent, not real, motion of the tail and whole body is also horizontal, instead of being vertical. There can be no doubt whatever, to the philosophical zoologist, that all the flat-fishes have been, in course of time, modified from the ordinary and more symmetrical type of fish. Even now,

we occasionally get what is termed a " monstrosity,"
in which a flat-fish has its eyes one on each side—
this being a reversion to the ancestral type. During
the earliest stages of all flat-fish within the egg they
are like ordinary fishes ; as they develop, one eye
may be seen gradually turning round to the same

Fig. 155.

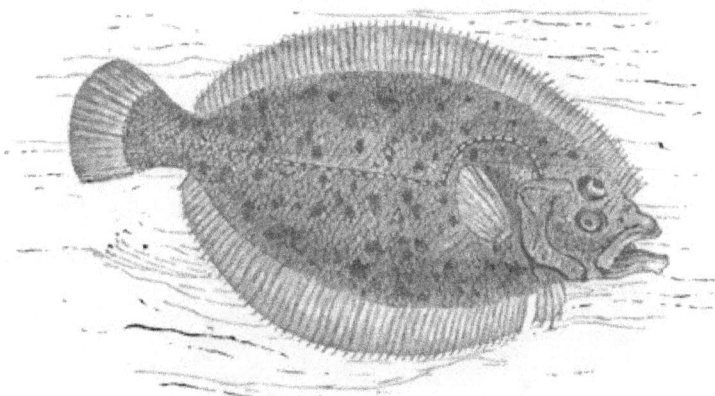

The Brill (*Rhombus lævis*).

side as the fixed one, and in this way we have a
flat-fish, modified to swim and rest on one side, and
perfectly adapted to live under what seem the most
singular conditions we can imagine of any animal!
The skull partakes of the same embryological modification, being symmetrical when the fish is in the
egg, and gradually changing to an unsymmetrical
state as the fish gets older. Some of the flat-fish
attain a great size, especially the halibut, turbot,
and brill ; whilst others, as the little dab (*Platessa*

Q

limanda), never acquire great size or weight.   This graceful fish must be seen alive in the tanks to be properly admired.   There are several species of dab,

Fig. 156.

The Dab (*Platessa limanda*).

but the above is the commonest.   The specific name of *limanda* is derived from the Latin word for a "file," in allusion to the roughness of the scales on the upper side.   All the flat-fishes live on small mollusca and crustacea, alive and dead.

In addition to the fishes mentioned in the last three chapters, which practical experience has proved may be more or less readily acclimatised in marine aquaria, there are many others, and the list is being added to almost every week. The conger-eel (*Conger vulgaris*); john dory (*Zeus faber*), a lovely fish when alive, whose common name is corrupted from the French *jaune dorée*, an apt allusion to its burnished golden body; the fork-beard (*Raniceps trifurcatus*); the singular gar or guard fishes (*Belone vulgaris*), whose slender elongated, silvery body terminates in formidable jaws, armed with sharp teeth; the mud fish—a singular illustration of the "missing links" between amphibians and fishes; the lovely and graceful smelts; sting rays; sur-mullets; skates of all kinds, &c., are among the commoner kinds exhibited alive. The study of marine fishes is now removed from the mere examination of dried skins or shrunk specimens preserved in spirits, to where they can be seen in their natural element, graceful as butterflies in their motions, and many of them hardly less brilliantly coloured. There we can watch out every stage of their life-history, from the extrusion of the spawn to the adult fish, and can understand from their habits of life the meaning of many a structural peculiarity, many a tint and spot and ornament, which before we should have rashly assigned to some freak of Almighty Power, unaware that we were then exercising a mental act that savours of blasphemy!

CHAPTER XIV.

CUTTLE-FISH, MOLLUSCA, ETC., OF MARINE AQUARIA.

IN all our large marine aquaria no object has been more popular than the octopus, or "devil fish," as it has been more emphatically called. The weird stories told of it by Victor Hugo, in his 'Toilers of the Sea,' had prepared the public mind for something so exceedingly ugly as to be unusually attractive ; and accordingly the first specimen of a living octopus in the Crystal Palace Aquarium had to bear the uninterrupted gaze of lookers-on for weeks. It sat for its portrait in the illustrated papers, and had all its points noted down by newspaper correspondents with the same faithful detail as if they were those of prize cattle at the Agricultural Show. Brighton afterwards became possessed of one of these animals, and fortunately Mr. Henry Lee was there to study its habits, and to embody them in a series of papers which were collected into a volume not long ago on 'The Devil Fish of Fiction and of Fact.' This is the most interesting work on the cephalopoda we have in our language.

Since Victor Hugo so largely drew upon his vivid

imagination in the description of his peculiar species
of cuttle - fish (or "cuddle"-fish, from the powers
of embracing possessed by its long arms), portions of
gigantic specimens have been found off the coasts of
Newfoundland, and described in the scientific journals.
These fragments indicate the actual, but fortunately
rare, existence of cuttle-fishes nearly 30 feet in length,
arms included. The old fishermen's stories of boats
being sometimes enveloped by the arms of these huge
" krakens," have a semblance of truth. The common
octopus (*Octopus vulgaris*) is now kept in all our marine aquaria. Its structure is very peculiar, for the
water admitted into a special chamber for aerating
purposes, can be so expelled as to subserve the
purpose of locomotion. The animal usually crawls
along the sea-floor head downwards, moving about by
means of its long tentacles. But when it wishes to
move more rapidly, all these are drawn together in front
of the head, the water is jerked out of the branchial
chamber through a special funnel, and the cuttle-fish
is thus driven backward by the rebound. The suckers
on the tentacles are most formidable organs for retaining hold, each one being provided with a natural
piston, so that a vacuum can be created when it is
withdrawn at the will of the animal. The horny
mandibles of the mouth are very much like those of
the parrot, and by their means the-cuttle fishes can
bite through the carapaces of the crabs, &c., on which
they habitually feed. Although this species pos-

sesses an ink-bag its contents are rarely poured forth ;
but when alarmed, the ink-bag is emptied suddenly,
and the water is then so beclouded with the inky

Fig. 157.

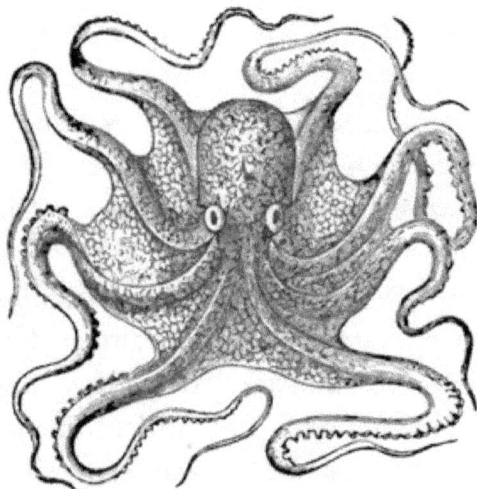

" Devil Fish " (*Octopus vulgaris*).

Fig. 158.

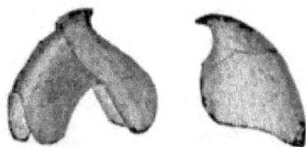

Horny jaws, or mandibles of Cuttle fish.

fluid that the cuttle-fish can make its secure escape
whilst the water is disturbed. The octopus has the
peculiar power of changing the tints of its skin, according to those of the ground it may be reposing

upon ; for its habits are not very active, and it will remain in the same position for hours, occasionally coiling and uncoiling one or another of its eight long sucker-clad feet or tentacles. A stock of live shore-

Fig. 159.

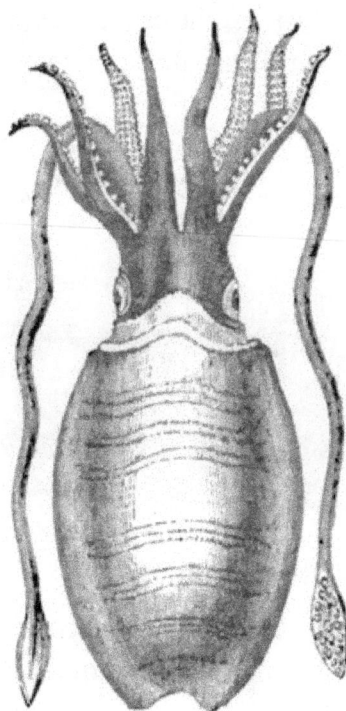

Fig. 160.

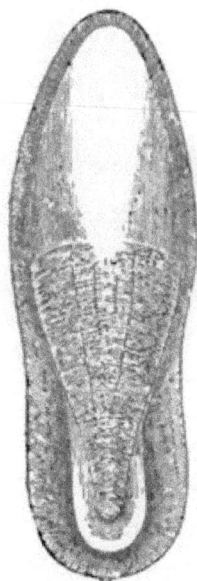

Common Sepia (*Sepia officinalis*).

Bone of Common Sepia (*Sepia officinalis*).

crabs (*Carcinus mænas*) is usually kept in a separate tank for the purpose of feeding cuttles.

The common sepia (*Sepia officinalis*) is also kept under artificial marine conditions, but it is not so great a favourite, either with the public or the managers, on account of the readiness with which it discharges the contents of the ink-bag, so as to discolour the water of the tank. This species is a decapod, not an octopod—that is, possesses ten feet, of which two are long and retractile, instead of eight, as in the octopus. The so-called internal "bone" of the *sepia* is a common object along our coasts, where

Fig. 161.   Fig. 162.

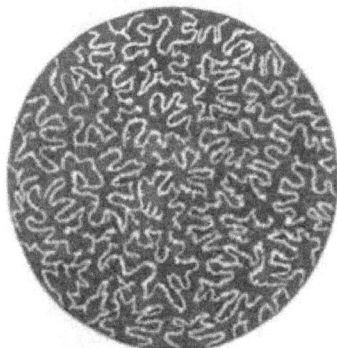

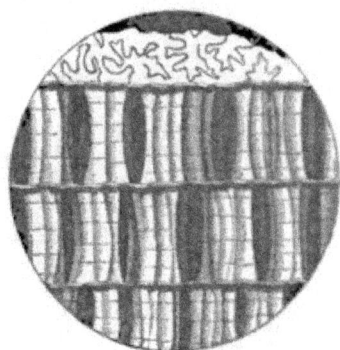

Sepiostaire (ground plan).   *Sepiostaire* (vertical section).
(Magnified.)   (Magnified.)

its whiteness soon attracts attention. Of course in this condition it represents a once living animal, just like any other skeleton does. These "bones" are termed " sepiostaire," and are collected and ground up for tooth-powder, the calcareous matter retaining its crystalline structure to the smallest particle, and thus being an admirable dentifrice. Slices of the sepiostaire, both vertically and horizontally, make beautiful objects for the microscope, and show the mode in which it is built up. The common squid (*Loligo vulgaris*) is also a decapod, two of its tentacles being much longer than the rest. Like the sepia and others of the same group, it possesses an internal bone or

Fig. 163.

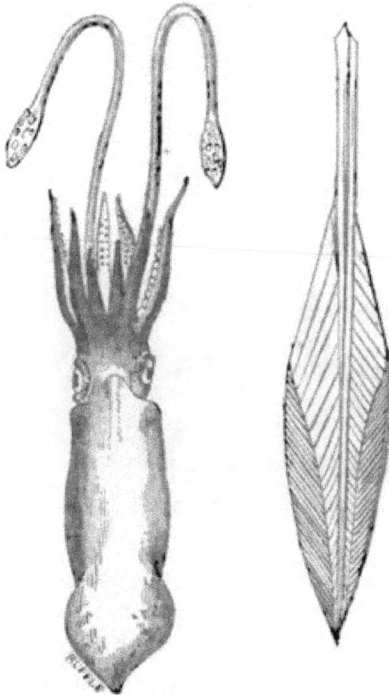

Common Squid   Pen of ditto.
(*Loligo vulgaris*).

"pen," the latter name being given to it on account of its semi-transparent horny or quill-like structure. This "pen" contains no limey matter. The *Sepiola Rondeletii* is a much smaller cuttle, and addicted to roving habits. It is not uncommon as an aquarium object, although it gives some trouble through its ink-discharging habits. Besides the above-mentioned species, another octopus (*Eledone cirrhosa*), having only one row of suckers on its feet, is kept in most of our public aquaria.

Fig. 164.

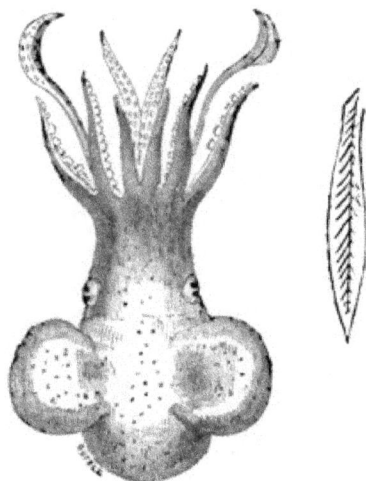

Sepiola Rondeletii.   Pen of ditto.

Although the cuttle-fishes and their allies are un-doubtedly at the head of the mollusca—a position warranted by their specialised and perfect organs of sight, brain (enclosed in a cartilaginous box, sug-gestive of a skull), mouth, stomach, locomotive powers, &c.—they belong to one of the most ancient groups, geologically speaking. We meet with some of their ancestors in the Cambrian strata ; although the recent type of cuttle-fish, with the hard parts internal instead of external, dates back no further, perhaps, than the oolitic epoch.

Shell fish, both univalve and bivalve, are kept in aquaria, according to their habits; although the limpet, chiton, fissurella or "key-hole limpet," dog-whelk, and periwinkle are peculiar to shallow waters,

Fig. 165.

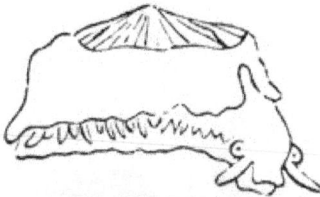

Animal of Fissurella.

Fig. 166.

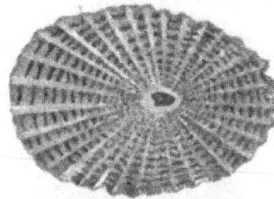

Key-hole Limpet
(*Fissurella Græca*).

and will creep out into the air. These are most of them easy of domestication in such shallow tanks as an amateur would commence with. All are in-teresting objects, the limpet family, including the several species of chiton, being particularly so. The latter looks like a woodlouse, and has the power of partially coiling itself up when detached from the rock, to which, however, it clings as long as it can with all the proverbial force of a limpet. Formerly it was called a "multivalve" shell, on account of this latter being made up of a number of transverse pieces. It is, however, a near rela-tive of the limpet. Their young are shell-less, and have a semi-ring of cilia around the margin of the upper half, by means of which they can swim about freely. The handsomest of our British chitons, per-haps, is *Chiton fascicularis*, so called on account of the bundles or tufts of bristles which crop out between the plates. This species crawls backwards and for-wards with equal facility. The key-hole limpet (*Fissurella Græca*) is tolerably abundant around the more southerly shores of Britain.

Fig. 168.

Fig. 167.

Chiton fascicularis.

Egg-cases of White Whelk.

The white whelk (*Buccinum undatum*) as well as the red whelk (*Trophon antiquum*) is a good aquarium object, active and vigorous in its movements; and may be useful for devouring any decomposing animal matter. The white whelk is usually caught by lower-ing pieces of decaying fish or flesh in shallow places; the whelks soon scent these out, and crowd over them in immense numbers, and are then caught by hoisting up the fragments to which they are attached. In the eastern counties these molluscs are largely eaten, and with good reason, for their flesh is so much like that of the edible cockle that it might easily replace it as a culinary object. The egg-cases of the whelk are very common objects by the seaside, of a light brown colour, resembling a large head of hops. Each capsule contains an individual embryonic shell, perfectly formed, and not exceeding a pin's head in size. When they are present in a storm-tossed specimen of egg-cases, they are detached, and rattle inside the capsules when shaken. *Haliotis tuberculata* or "Venus' ear," is another marine mollusc of much service in the aquarium as a vegetable scavenger. It is lively and attractive both on account of the graceful shape of the animal, and the prismatic tints which adorn the shell. *Nassa purpura* (or dog-whelks), top shells (*Trochi*), murex, *Pileopsis, Cyprea, Aporrhais pespelicani* (or pelican's foot shell), *Natica monilifera* —a common and prettily marked univalve shell, easily domesticated—*Turitella, Bulla,* &c., are other univalves which are kept with more or less ease in the Brighton, Crystal Palace, and Manchester Aquaria.

The group of shell-less molluscs—answering in the sea to the slugs upon land—have been favourites in the marine aquarium ever since Alder and Hancock, and especially Gosse, wrote so attractively about them twenty-five years ago. But the sea-slugs breathe by means of gills, which are usually borne on the back, uncovered by any shell, hence the name given to the group of *Nudibranchiata,* or "naked-gilled." These branchiæ or gills are usually coloured, or prismatically tinted, so as to render their possessors

very beautiful and attractive objects. In another
group (*Tectibranchiata*) commonly to be seen in marine tanks is the *Aplysia*, or "sea-hare"—so called
on account of the rude resemblance of the head and
shoulders to those of a "sitting hare." The *Doris*, or

Fig. 169.

Sea-slug (*Æolis coronata*).

"sea-lemon," well deserves its popular name both
from its colour and general appearance. These sea-
slugs usually construct very pretty "nidal-ribbons," or
egg-cases, in which the eggs are placed in thick rows,
like beads closely stitched over the surface of a piece
of ribbon.

Among the bivalves, we cannot wonder that the
oyster and the mussel are favourites. The latter is
abundantly kept in the tanks, on account of its enormous powers of reproduction. The issue from a
single living specimen would soon multiply at such a
rate as to clothe the entire surface with dense masses
of mussels ; but the young are greedily sought after
by many fishes, &c., so that they are advantageously

placed in tanks where their reproductive powers can
be turned to account in feeding the other objects.
We have several British species of mussel, the prettiest

Fig. 170.

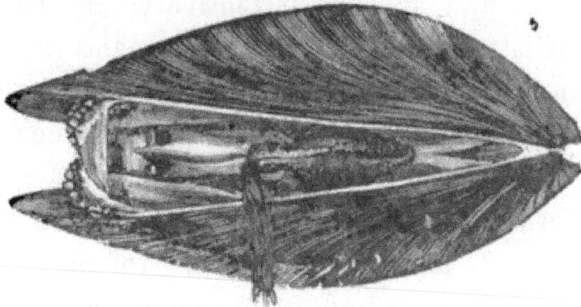

Common Mussel (*Mytilus edulis*).

Fig. 171.

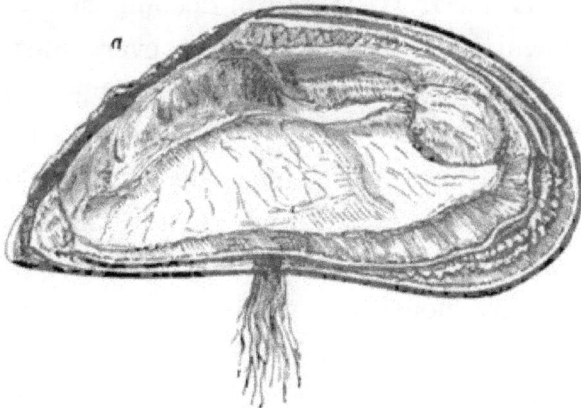

Common Mussel, laid open to show the plated gills.

of which is that called the "tulip mussel," a variety
of the common mussel, on account of the petal-like
stripes of colour which streak the shells. The com-

mon mussel (*Mytilus edulis*) is one of the most fertile
of all mollusca, and this natural history fact has
been turned to greater practical account off the French
coasts than off our own.   It is capital food, especially
when fresh, and as such is cultivated off the Nor-
mandy coasts, as we cultivate oysters, and thence imported into the interior in immense quantities as food
for the working people.   Off the Norfolk coasts the
mussel banks which form there so rapidly are used
chiefly for *manuring* the land.   The so-called "moss"
is in reality the *byssus*, or anchoring threads, and has
nothing poisonous about it, although this part has
been credited with producing the painful affection
known as "musselling."   We believe, however, that
the latter is due to partaking of decomposing mussels,
some of which are apt to get into every measure of
them, from the way some of the fishmongers have of
mixing up their old stock with the new.   In this way
a very old and rotten specimen may be made to taint
and infect all the rest during the process of boiling
and preparing for the table.

Oysters are never likely to lose their interest to
people who care anything at all for the pleasures
of the table; and our aquaria have already contributed some important facts concerning oyster
culture.   Apart from their well-known and highly-
esteemed edible qualities, they are favourites with
managers of large marine aquaria from their useful-
ness in *clearing* turbid sea water, and rendering it

transparent. Mr. Lloyd believes this is due to their breathing out quantities of carbonic acid gas, and thus converting the carbonate of lime which gave the water a milky appearance, into a *bi*-carbonate, when the visible carbonate is taken into suspension and thus rendered invisible. Oysters are therefore to be seen distributed in nearly all the Brighton tanks, on account of the temporary service they render in keeping the water transparent. Their semi-opened shells exhibit the plates of gills (commonly called the "beard" of the oyster). These gills are richly clothed with active cilia, which create currents in the water, and thus the gills are constantly bathed by fresh supplies of oxygen, the same currents also bringing food.

Fig. 172.

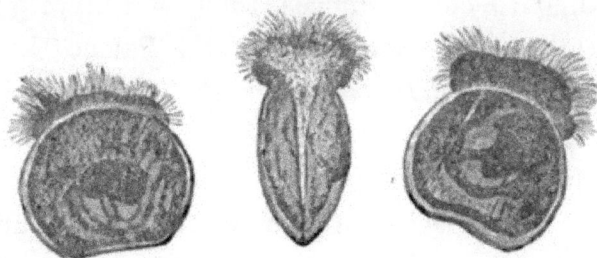

Young Oysters, with their natatory or swimming gills.

A good deal of the turbidity of sea water in tanks is often due to the presence of myriads of the zoospores of algæ. These zoospores have a wriggling motion, so that few people uninstructed in the life history of sea-weeds, would imagine them to be connected with the reproductive parts of maritime

R

plants.   No doubt these zoospores form a good part
of the food of such sedentary mollusca as oysters,
mussels, &c.

The young of the oyster do not leave the folds of
the mother until they are capable of moving about
and seeking their own food.   At first, therefore, oysters
are free-swimming, moving about by means of special
tufts of cilia.   A single oyster will throw out myriads
of such embryos, which go by the name of "spat."
In many respects these very young oysters resemble
the lower crustacea, the *Entomostraca*, for like them,
they have two shells, through which the natatory
gills are protruded.   When the young oysters settle
down to the same staid life as their parents, the spat
is then said to "fall."   In from one to three years,
according to circumstances, they will have reached
the adult stage.   At the Crystal Palace the American clam (*Cyprina moneta*) is shown alive.   Indeed,
there are few bivalves which cannot be healthily
kept in aquaria, especially if the conditions are
right.   Two species of cockle, the edible and the
spiny (*Cardium edule* and *C. echinatum*), may usually
be seen in tanks.   The burrowing and leaping habits
of the cockles are very interesting to witness.   The life
of the young of the common species is very much like
that of the oyster.   The recurved spines of *Cardium
echinatum* are bent in a direction contrary to that in
which the animal burrows, so that they do not impede
its habits.   All the species of scallop-shells (*Pecten*)

move by means of alternately opening and closing the flattish shells with a kind of jerk. This enables them to dart backwards to some distance ; and, as some of our British species have very richly coloured shells (especially those obtained off the south-eastern coasts), it follows that they have a butterfly look when seen in the water. Around the edge of the mantle, through the partly opened shells, may be seen a row of eyes, looking like diamonds set in a ring, except that the latter can give no idea whatever of the depth of colour and play of light which these eye-spots of the *Pecten* seem to contain.

Fig. 173.

Scallop (*Pecten*).

The burrowing powers of some bivalves is not limited to sand or mud. We find them able to excavate a protection for their fragile shells in the hardest rocks ; limestones, however, being evidently preferred to any other. The *Pholas, Saxicava, Teredos,* and others, are able, nevertheless, by means of the rough thick part of these shells, near the hinge, to mechanically excavate the holes in which we find such mollusca living. The mode in which these holes in rock were formed was discovered by keeping pholades in an aquarium, by Mr. Robertson of Brighton, who carefully watched the entire process and wrote a detailed account of it. These holes are

R 2

always widest at the bottom, and narrow at the top, which communicates by means of a fleshy syphon or tube with the external water.

Fig. 176.

Fig. 175.

Fig. 174.

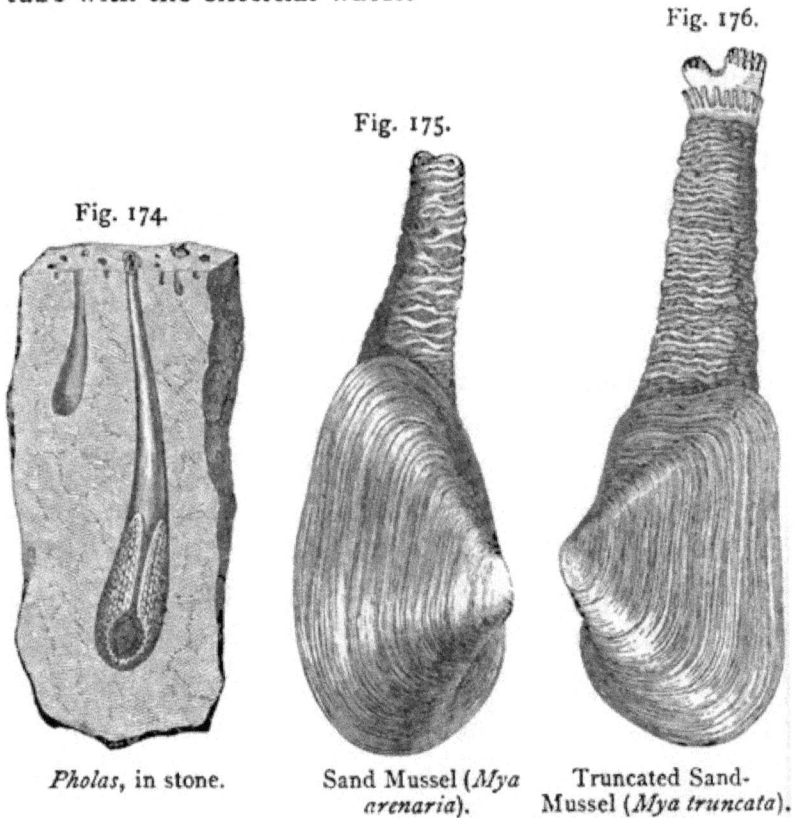

Pholas, in stone.   Sand Mussel (*Mya*   Truncated Sand-
arenaria).   Mussel (*Mya truncata*).

In some bivalves these syphons are double, as in the sand mussels (*Mya arenaria* and *M. truncata*) ; a current of fresh sea-water goes down one, and the water, after passing the gills and mantle, and being partly deprived of its organic matter and oxygen, is returned up the other. These mussels burrow in mud and sand, so that little is seen of them except the tips

of their coalesced syphons, which are covered with a
fringe of cilia that guard against the entrance of
unsuitable matter. Both species of these common but
elegant shells are kept in our public aquaria. In addition to the stone-borers (*Saxicava* and *Pholas*) at
the Crystal Palace may be seen the wood-boring
mollusca (*Xylophaga* and *Teredo*). The latter goes
by the common name of "ship worm," but it is in

Fig. 177.

Mactra stultorum.

reality a mollusc in which the shells are reduced to a
minimum of size, the lime secreted by the animal
being used to line the winding worm-like burrows it
makes in the wood where it takes up its abode. The
Mactras, Tellinas, Donax, and *Venus* are other prettily-
coloured and elegant aquarium mollusca. *Cyprina
islandica, Modiola, Modiolus* (the latter usually called
the "horse-mussel"), *Pinna pectinata, Lima hians,
Scrobicularia piperata,* &c., are many of them larger

in size, but all have been or are kept in captivity at
the Crystal Palace. At Brighton, some of them, as
well as species of echinoderms, appear to be kept with
great difficulty, or cannot be preserved at all.

Fig. 178.

Donax politus.

Around the margins of the rockwork in most of the
large marine tanks may be seen rows of greenish-
white objects, not unlike elongated white Hamboro'
grapes, which are fixed by their bases.   These are the
Ascidia, a very interesting group of mollusca allied
to the polyzoa, which some naturalists have actually
included among the *Vertebrate* animals.   For reasons
which are partly structural and embryonic, they are
usually regarded, however, as nearly related to the
mollusca.   They are in reality *soft-coated* mollusca,
just as oysters and others are hard-coated.   Four
species of these ascidians, *Ascidia intestinalis, A.
vitrea, Molgula tubulosa,* and *Cynthia quadrangularis,*
have semi-spontaneously made their appearance in

the marine tanks of the Crystal Palace, Brighton,
and elsewhere. They are pretty objects, increasing
at an abundant rate. The embryos of some of them
(notably the *Cynthia*, commonly known as the "cur-
rant squirter") are free-swimmers, and have a peculiar

Fig. 179.

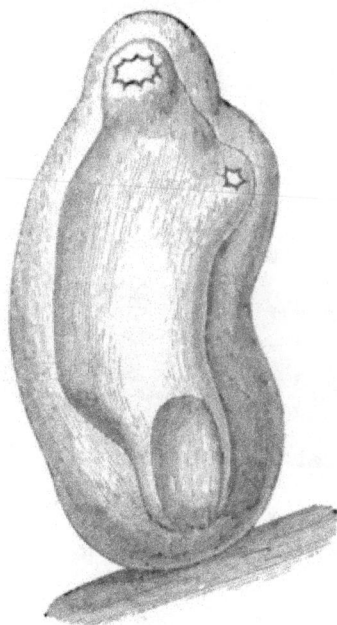

Ascidia mentula.

Fig. 180.

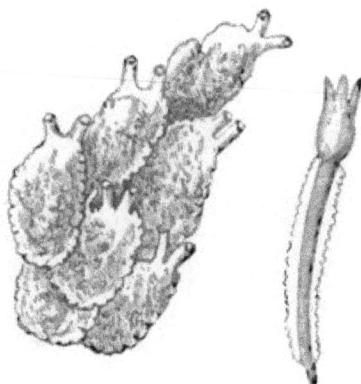

Cynthia, and its Tadpole.

tadpole-like appearance. Among fishermen all the
ascidians go by the popular name of "sea-squirts,"
from the ease with which they can eject a jet of
water when their leathery outer tunic contracts;
hence their name of *Tunicata.* These peculiar, and
to the philosophical naturalist most interesting,

animals, are not distantly related to other marine
objects called "sea-mats," whose dry brown fronds
are frequently to be picked up along the sea-coasts,

Fig. 181.

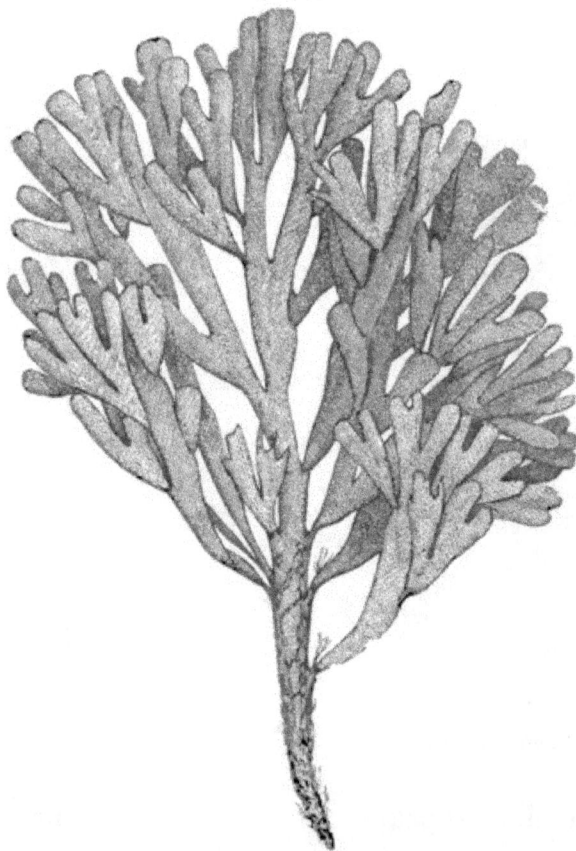

Sea-mat (*Flustra truncata*).

where they are usually mistaken for sea-weeds, and
mounted as such in albums. They are, however,
colonies of really highly organised individuals, of

microscopic smallness, which live as neighbours in the horny frond they have secreted in common A magnifying glass will show that the surface of such frond is covered with cells, all shaped alike. In these the polyzoans live, obtaining fresh air and food by means of their cilia, which are protruded so as to be constantly agitating the water. These sea-mats may be kept alive in small aquaria, when even by the naked eye we can perceive the extrusion of the cilia by the filmy bluish-whiteness which seems to come over the surface of the frond. Each individual of the colony forming the sea-mat lives separately from the rest. In this respect,

Fig. 182.

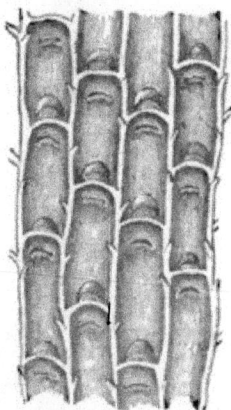

Magnified portion of *Flustra.*

therefore, they are utterly unlike the *Sertularians,* or "sea-firs," in which the individual polpes are connected by a common flesh which runs up the horny stem, and is given off to every branch so as to be connected with every zoophyte. Moreover, the animals forming the sea-mats have a nervous system, which the sea-firs have not, as well as a more complex physiological organisation. Some of the members of this family may be seen encrusting sea-weeds with a most delicate white lace-like tracery — the lace-work being produced by the cells of the animals. The commonest of these is *Membranipora pilosa,* which may be found wherever sea-weeds are to be gathered, clustering their stems or spread out over the fronds. It is held by naturalists that the appearance of sea-mats, sea-squirts, &c., in aquaria,

Fig. 183.

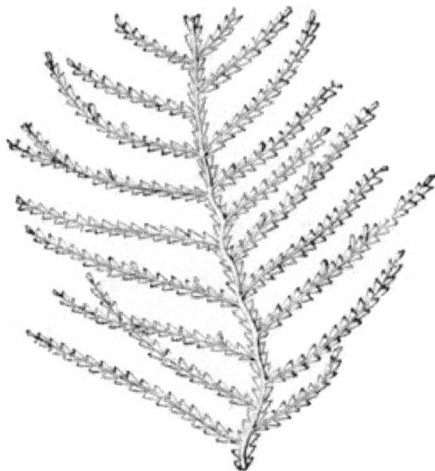

The Sea-fir (*Sertularia abietina*).

is a proof that the water is in a healthy condition. Another allied and abundant colony of marine objects—all of them having affinities with the mollusca in spite of their apparent differences—may be seen on the backs of the larger sea-weeds and other objects. They look like a firm layer of hardish jelly, in which are scattered "stars." We have several genera and species of these pretty creatures, all of which are apt to make their appearance in healthy tanks. Some are joined at their bases by means of a thread-

like connection (*Clavelina*); but those in which the
individuals form star-like masses go by the name of
Botryllus. This really signifies a bunch of grapes,

Fig. 184.

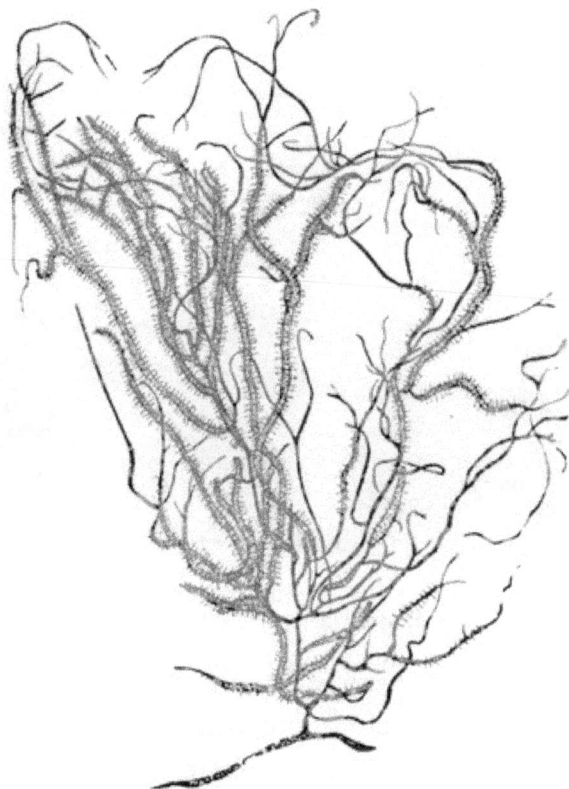

Membranipora pilosa, encrusting sea-weed.

and is an allusion to the way in which the individual
members of a *Botryllus* colony are related to each
other, just as the grapes on the same bunch are.
Each "ray" of a "star" (Fig. 186) is a separate

animal; but all "are arranged in this stellate form because they have one *anus* in common, out of which excreta, &c., are ejected." The mouth of each animal may be seen like a dot in the "ray."* This is sur-

Fig. 185.

Fig. 186.

Botryllus on frond of sea-weed.

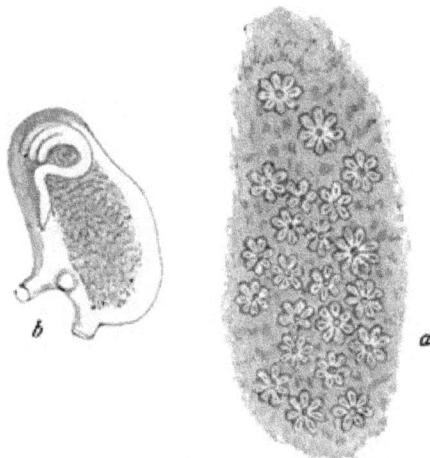

a, *Botryllus polyclas*; b, Separated individual, forming a "ray" of one of the star-like objects.

rounded by cilia, and when alive, is in reality a very active organ, producing miniature whirlpools in the sea water, and thus obtaining fresh oxygen and food. The common anus forms the centre dot of each

* Throughout this book such *circulating currents* in various animals are frequently mentioned, and they are now particularly adverted to, because it is by imitating them in the machinery of aquaria that the best biological results are attained. In this way, perhaps unconsciously, we do as Nature does, and then find that she, long previously, has done the same thing in a far neater and better manner.

"star." We have long since learned that the mere magnitudes and even shapes of animals are not the most important characters by which to associate groups, but this fact is nowhere felt in such a degree as when we study the polyzoa and their allies.

CHAPTER XV.

CRUSTACEA, ECHINODERMS, ANNELIDS, ETC., OF MARINE AQUARIA.

ONE of the most important advantages which large aquaria, aerated by constant jets of water, possess, is the great variety of marine animals, vertebrate and invertebrate, which may be kept in them. Not only may they be so selected as that one group shall not harm another, but much of the labour and anxiety likely to occur from the pollution of the water by decomposing food or dead animals, may be prevented by merely including certain omnivorous creatures which will clear away such garbage and consume it as food. Some of the carnivorous mollusca (the whelks, for instance) are useful in this respect, but many of the crustacea are even more so. The latter, also, are more lively animals, and never fail to cause amusement by their grotesque and serio-comic habits. One cannot witness the rude gambols of lobsters and crabs without feeling that the element of humour is not merely a subjective condition of the human mind, but has an objective existence in nature.

Of all the useful and interesting scavengers, none is more so than the hermit crab (*Pagurus Bernhardus*), so called from its occupying an empty shell, as a hermit would his cave. This crustacean is forced to this singular habit by the softness of its abdomen, which does not secrete a hard crust, and therefore

Fig. 187.

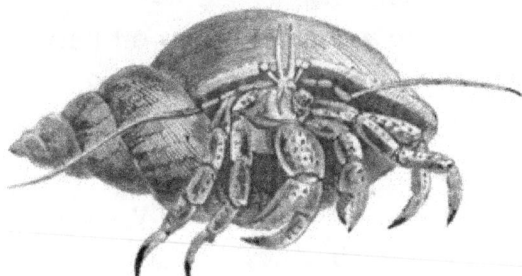

Hermit Crab (*Pagurus Bernhardus*).

requires such protection as an empty whelk-shell will afford. Nothing on earth exceeds the shamefacedness of a hermit crab deprived of its shell—not even a bather whose clothes have been stolen! When perfectly accommodated with an empty shell, the hermit crab is most suspicious and wary. It regards everything as an enemy; and even when one of its own kind approaches, you will see it move away, or draw itself within its cave, and close the aperture by means of its large right claw, which is bigger than the other for the purpose. Being as warlike as they are suspicious, a good many fights come off in the tanks between them, insomuch that this species is also called the "soldier crab." The somewhat vulgar practice is now followed of introducing highly-coloured empty tropical shells for the hermit crabs to take up their abodes in, although this has added to the attractive appearance of the bottoms of the tanks. Before moulting these crabs generally leave their old domiciles, and select one so much bigger that they can move about in it. Their increase in bulk is then usually very rapid. Many a contest comes off between hermit crabs, when two of them wish for the same empty shell. Every atom of food rejected by other animals in the tank, and which would otherwise lie on the floor and foul the water, is greedily cleared away by hermit crabs and their allies.

The graceful appearance of lobsters when at rest makes them prominent objects. They seem to be in almost a devotional attitude, resting on their huge pincers as the " praying mantis " does on its fore-legs ; and in this position are often seen with their faces towards the glass front. Their long, slender, jointed antennæ are thrown backwards, and are in a constant state of motion. The spiny lobster, or " sea-crayfish " (*Palinurus quadricornis*), is a more attractive marine object than the common lobster, although its flesh is not such delicate food as that of the latter. Its body is covered all over with spines and prickles, and is moreover very brightly coloured. In length this species exceeds any other British crustacea. It is noticeable, however, by the absence of the large and powerful pincers which distinguish the common lobster. The females of this species have spawned both

Fig. 188.

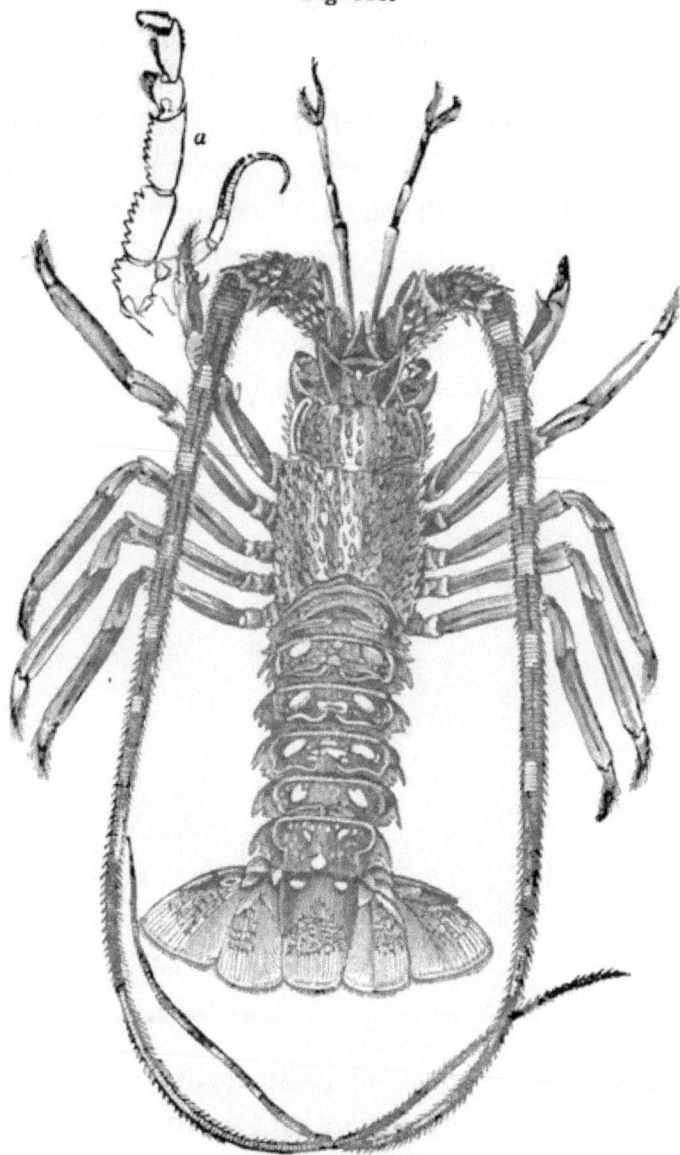

Spiny Lobster, or Sea Cray-fish (*Palinurus quadricornis*).
a. Left outward foot-jaw.

at Hamburg, Brighton, and the Crystal Palace, and the
tanks were then crowded with transparent, leaf-like
young, which, before then, were regarded as a distinct
species, and called the "glass crab" (*Phyllosoma*).
The common lobster (*Homarus vulgaris*) is an aquarium favourite, and although its colours are not so
bright as those of the foregoing, the plum-tinted
carapace is not without beauty; whilst its graceful
motions in walking and climbing by means of its
slender feet, and swimming either by the aid of the

Fig. 189.

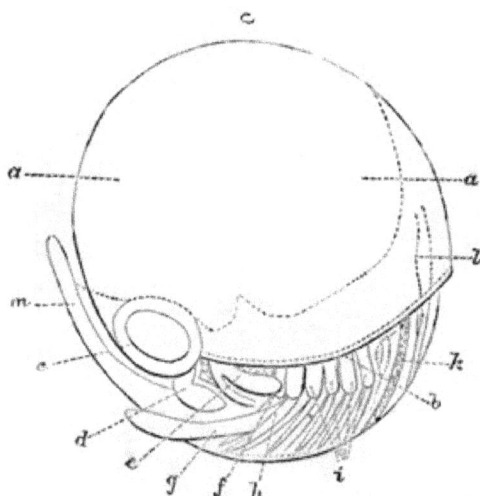

Embryo of common Lobster, magnified 20 diameters.

"swimmerets" arranged underneath the abdomen as
so many fringed plates, or by one flap of the powerful
expanded, fan-like divisions of the tail, give a good
deal of animation to a tank. When several of these

crustacea are together the interest is much increased. The development of the young of the lobster, from the numerous eggs which the female usually carries under her body in dense masses, has recently obtained a good deal of attention. It is most interesting as indicating the *nauplius* stage, which characterises the early embryos of nearly all crustaceans alike, no matter what their adult differentiations may be. The young of the lobster pass through several stages

Fig. 190.

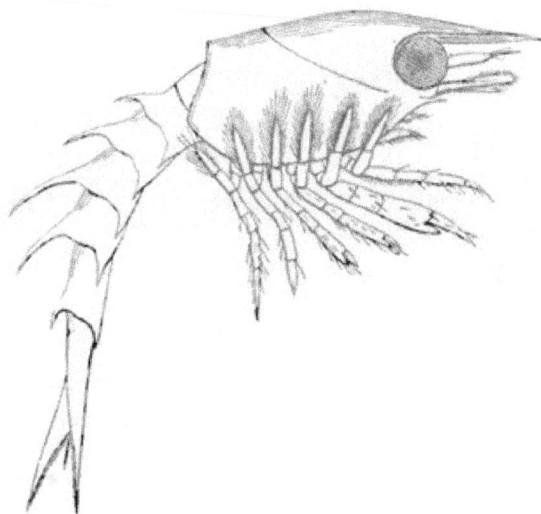

Larval or mysis stage of the common Lobster.

before they reach the adult condition. The first is visible before the egg is hatched. In this condition the carapace (*b*) is indicated by the presence of spots of red pigment, the rudiments of the eye (*c*), antennæ

S 2

(*d* and *e*), of the great claws (*g*), and the bilobed tail (*m*), are plainly visible, as well as the most important of the internal organs, such as the heart (*l*), intestines (*k*), &c. On hatching, the second condition, called the *mysis* stage — because it then resembles the adult condition of the opossum shrimp, or *Mysis*—is next undergone. The young lobster is now about one-third of an inch in length, and, as will be seen in Fig. 190, possesses six pair of legs, one pair being subsequently modified into footjaws. In the third stage, when the larva has attained a length of about half an inch, it loses its *mysis*-like appearance, and begins to assume something like its adult features. In the *mysis* condition it swims on or near the surface of the water ; and even in the next stage is more or less of a free swimmer, these habits not being left off until after several succeeding stages of its development. At present it will be seen from Fig. 191 that

Fig. 191.

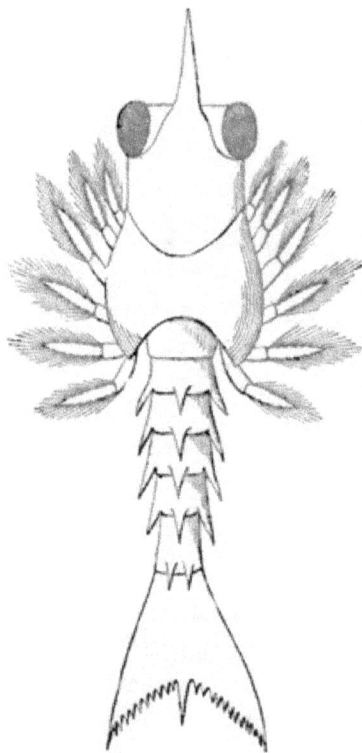

Back view of fig. 190.

rudimentary " swimmerets " have appeared on the
second to the fifth segments of the abdomen, whilst
the large claws are in process of formation. Even
after the young have reached what is called the *adult*
stage, they are so unlike fully developed lobsters that
they might be regarded as a different genus. Their
movements are now very much like those of shrimps,
and they frequent the surface of the water much more
than the bottom. All the above changes are believed
to take place in a single season.

Fig. 192.

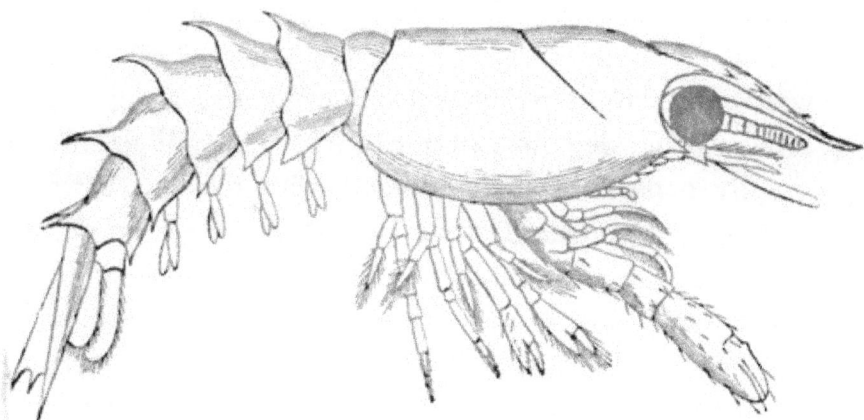

Third larval stage of Lobster.

Like all the crustaceans, thick-tested species particularly, the lobster increases in size by moulting or
casting off its old coat, which is thrown off in one
piece, and looks so perfect that it might be taken for
the animal itself. Not unfrequently it casts an odd

limb, or has one torn off in a fight. This difficulty is got over by a new one budding. Mr. Lloyd speaks of the general moulting as follows in his capital little ' Handbook to the Crystal Palace Aquarium ' :

" This moulting is necessary because the shell once formed never grows larger, and therefore as the creature within increases in size, and a new shell in a soft state begins to be formed below the old one, the latter becomes too small. The lobster is aware of this, and of its approaching moult, and instinctively knowing its utter helplessness from the attacks of its fellows, or from other animals during the quarter or half hour occupied by the disengagement of its shell, and while it is more or less soft for a few days afterwards, it, in an aquarium, sets about making a regular fortress, choosing its position with great judgment, usually beneath a shelving rock, with rock on each side, and with a kind of " earthwork " thrown up defensively in front, composed of the sand and shingle which have been removed from the hole in which it burrows, and here, ever on the watch for intruders, it patiently awaits its change of coat. Occasionally, when this fortification cannot be made, the lobster seeks a less perfectly protected place on a plateau of rock close to the water's surface, and therefore not often visited by other animals. When the time at last comes, it throws itself on its side, and ruptures the skin connecting the body with the first ring of the abdomen, and this is the only part intentionally

broken, though occasionally a limb is accidentally
separated in the operation. The greatest difficulty
seems to be in drawing the large anterior claws
through the comparatively very small dimensions of
the same limbs where they join the body, as, contrary
to what has been stated in books, these large claws
are not split open to allow of the emergence of the
limbs. The claws are soft, however, and are temporarily rendered shapeless by being pulled through
these little orifices. After the great limbs are free,
the rest is more easy, and by a series of spasmodic
efforts the remainder of the legs are extricated, together with the antennæ, great and small, and the
whole of the complex organs surrounding the mouth,
and even the eyes and the breathing organs are withdrawn from their old coverings, and while this goes
on, the tail is released. All proceeds simultaneously,
so that while one part of the process is being watched,
another is effected unobserved. When everything is
at last free, the lobster lies as if dead, and occasionally
does die from exhaustion, but generally it slowly
turns over on its legs, which, being soft, cannot support the body, however; but by remaining quiet, the
creature gains strength, hardness, and courage, and
the first use it makes of its returning vigour is to
thrust the old covering outside of its den, or else bury
it. But when a cast-off suit of armour can be secured
whole, it is about one-fourth less than the size of the
lobster which came out of it, and one can hardly

credit that it could have occupied so small a space. In about three days the newly attired lobster can go about with its mates on equal terms."

Among many other long-bodied crustacea (*Macroura*) which are acclimatised in marine aquaria, shrimps and prawns are of course very abundant. The former is the stock-food for an enormous number of fishes and other animals; and the intense

Fig. 193.

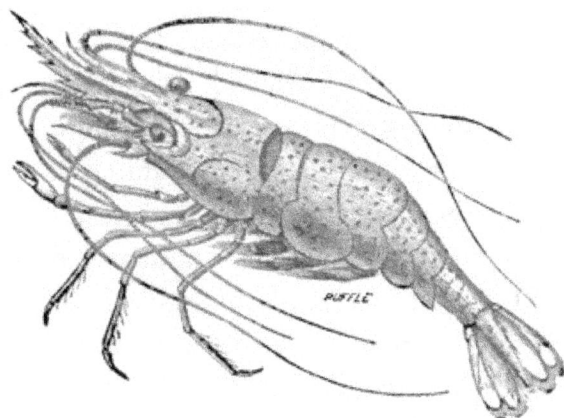

Prawn (*Palæmon serratus*).

stir there is created in a tank containing fishes when a supply of fresh shrimps is introduced is quite exciting. Both prawns and shrimps are hatched from the eggs, like crabs, in what is called the "*zoea* stage," in which the five hinder pairs of decapodal legs are wanting, whilst the two pairs of foot-jaws are employed as locomotive organs. These useful aquarium crustaceans may generally be seen

distributed through all the large tanks alike, but their
transparent appearance makes them difficult to be
seen. The usual position of prawns when at rest
is near the edges of rockwork, over which they glide
by means of their bristle-like feet. The Crystal

Fig. 194.

Common Shrimp (*Crangon vulgaris*).

Fig. 195.

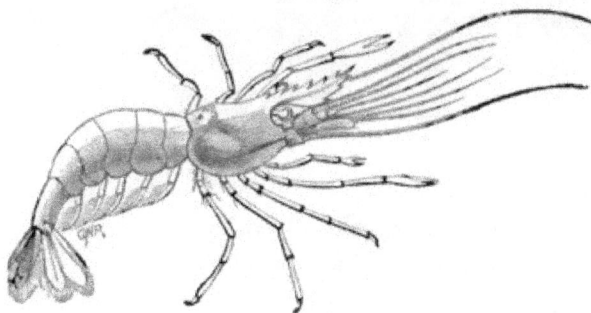

Palæmon squilla.

Palace Aquarium is especially rich in crustacea, both
long-tailed, as lobsters, shrimps (*Macroura*); and
short-tailed (*Brachyura*), as crabs. The shrimps love
to burrow and hide in the fine sand at the bottom,
and to throw a thin cloud of it over them by means

of their feet. When thus hiding, the speckled light grey colour of their bodies is admirably adjusted to the appearance of the floor. Among other aquarian crustaceans, some of them lower in organisation than those we have just mentioned, and others of which are more or less nearly related, may be mentioned the opossum shrimp (*Mysis;* the *Hippolyte,* an animal which assumes a bright green and

Fig. 196.

Fig. 197.

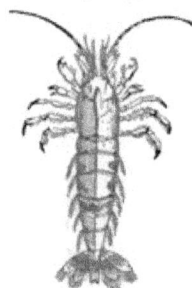

Hippolyte varians.   Banded Shrimp (*Crangon fasciatus*).

red colour at will; the banded shrimp (*Crangon fasciatus*); and the night-walker (*Nika edulis*). The banded shrimp is occasionally sold among the ofttimes miscellaneous mixture which goes by the name of "shrimps"; the night-walker has derived its name from its peculiar nocturnal habits. In addition to these may be seen the interesting *Gammarus* (also frequently sold and eaten as a "cup-shrimp"). No hen could exceed in anxiety for her chicks, the maternal

devotion of this little crustacean for her brood ; which
repay her for her attention by following her about

Fig. 198.

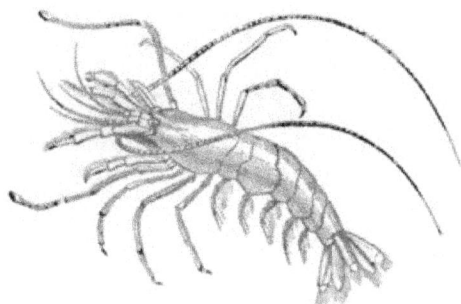

Night-walker (*Nika edulis*).

Fig. 199.

Gammarus locusta, and her brood.

after chicken fashion. Squat lobsters is the name
given to a group of brilliantly coloured crustacea

which, in some points, are intermediate between the
long-bodied lobster family, and the short-bodied crabs.
They are generally nocturnal in their habits, when
they are very stirring. By day they lurk beneath

Fig. 200.

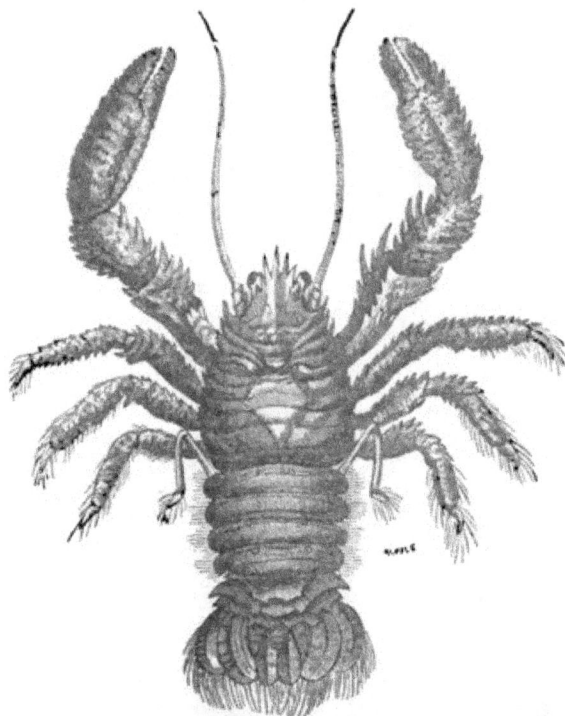

Squat Lobster (*Galathea squamifera*).

stones, &c., with their long tails folded up beneath
them, and in their natural habitats affect deepish
water. Off the Devonshire and Cornish coasts they
are not uncommon.

Of crabs we have of course an abundance. The

common shore-crab (*Carcinus mœnas*) we have already seen is, like the shrimp, regarded as "live stock" by aquarium managers, and generally kept in special tanks until required, or provisionally used as a scavenger, to clear away any docomposing animal matter. The edible crab (*Cancer pagurus*) is too valuable to be thus utilised, especially when the common, and to us uncatable "shore crab" will do as well. The spiny spider crab (*Maia squinado*) grows to a large size, and is as attractive among the short-bodied crustacea as the spiny lobster is among the long-bodied. All the spider crabs are nocturnal in their habits, and are generally so sluggish that we may see miniature forests of zoophytes and sea-weeds growing on their carapaces, or shells as they are popularly termed. It has been proved by Mr. Lloyd and others that one species, at least, is in the habit of daintily decking itself with sea-weeds, &c., as had been stated, and afterwards controverted. Miss G. Stephens has also recorded the fact as occurring in the genus *Pisa*. The body of the last species is covered with tough prickles. Nearly allied to it is the true spider crab (*Hyas araneus*), one of the largest of British species, if length of limb is to be taken into consideration in measurement ; these long legs having obtained for it the adjectival denomination of "spider crab." One allied British species is the four-horned spider crab (*Pisa tetraodon*), whose stout carapace is covered, especially along the margin, with strong incurved thorns. *Inachus Dorsettensis, Achæus cranchii,* and *Stenorynchus phalangium* are still better entitled to the name of "spider" crabs, on account of their small bodies and extremely long legs.

Fig. 201.

The Spider Crab (*Ilyas arancus*).

In addition to the above, the interesting group of "swimming crabs" has been introduced into the Crystal Palace and other aquaria, where at least half-a-dozen species are to be seen. The principal general are *Polybius, Portunus,* and *Portumnus.* Their adaptation to swimming habits is at once evident on seeing their flattened, oar-like hind legs. When swimming they usually take to the mid-water. The "masked crab" is to be found in some of our public acquaria, and is undoubtedly one of the most attractive. The male and female differ so much in general appearance that they

might easily be mistaken for different species. The
name is derived from a supposed resemblance which
the carapace bears to the human face. On some
coasts it is very abundant, and those acquainted with

Fig. 202.

Four-horned Spider Crab (*Pisa tetraodon*), male.

its habits can easily find it at low water, by slight de-
pressions in the sand, through which its long antennæ
are seen protruding. As a burrowing crab, these long
antennæ are of great service to it, inasmuch as they
are hollow, and thus form tubes through which the
water passes to the gills. The male especially has a
peculiar habit of occasionally sitting upright, like a
dog in the act of begging, and in this position its very
long fore-legs assist it materially. The "northern"
stone crab (*Lithodes maia*) is also a good aquarium

Fig. 203.

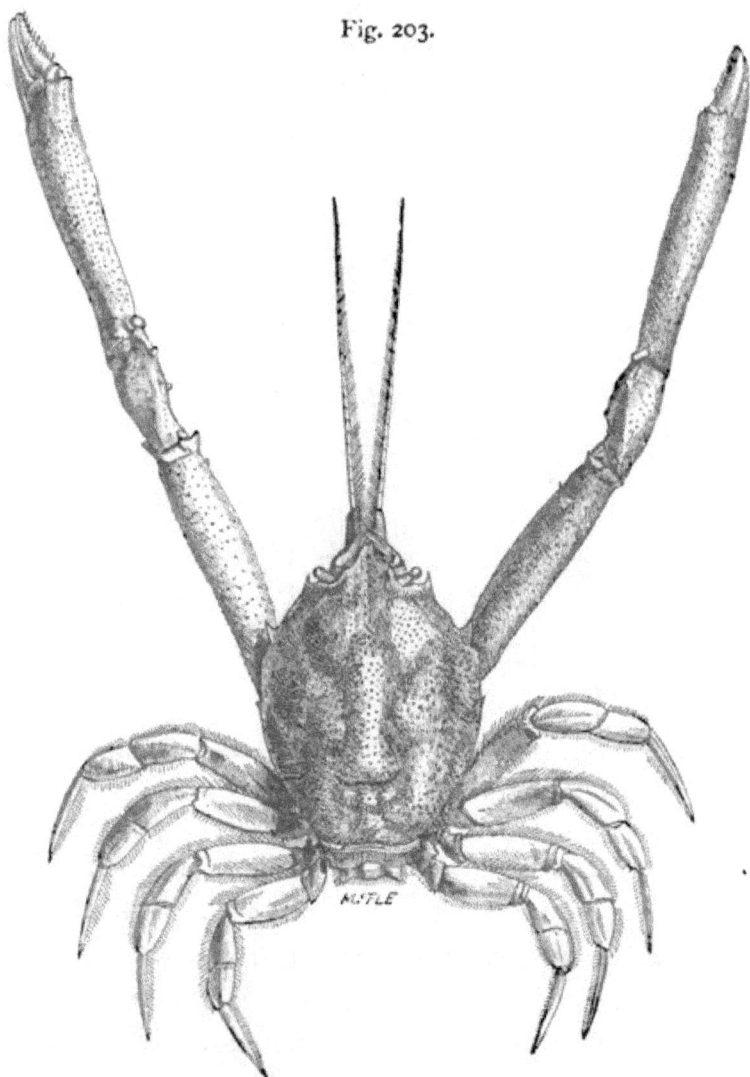

The Masked Crab (*Corystes Cassivelaunus*), male.

object, obtained abundantly on the Norway coasts,
although there was some difficulty in obtaining it at

first, owing to the superstition of the Norwegian sailors, who regard it as "bewitched." It is also found on the northern shores of Britain.

Fig. 204.

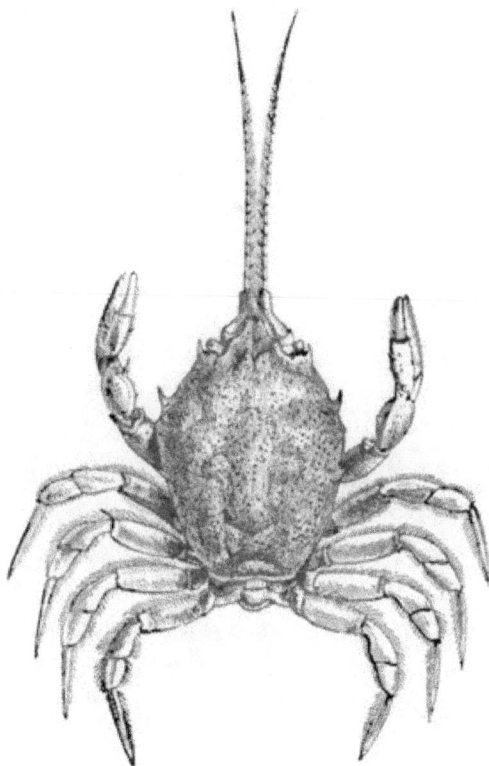

Female of Masked Crab (*Corystes Cassivelaunus*).

The Crystal Palace Aquarium possesses living specimens of *Euryonome aspersa*, *Hyas coarctus*, *Pirimela denticulata*, *Pilumnus hirtellus*, as well as the "nut" crabs (*Ebalia*), and the "angled" crabs

T

(*Gonoplax*). The former has the singular habit when at rest of simulating the appearance of rounded quartz pebbles. In addition to these may also be seen the interesting *Xantho florida* and *X. rivulosa*, as

Fig. 205.

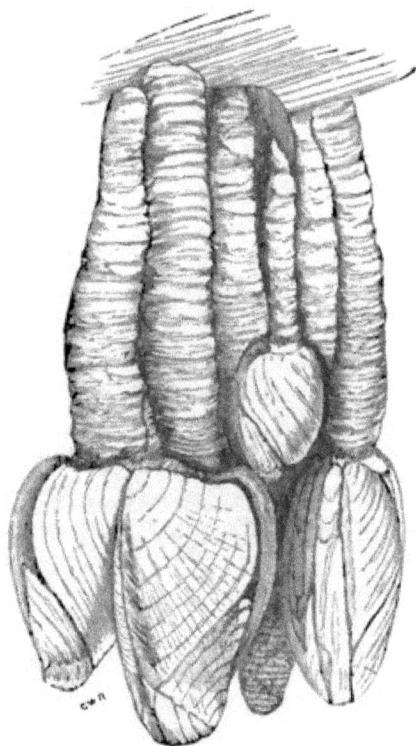

Stalked Barnacle (*Lepas anatifera*).

well as *Dromia vulgaris*, or "toad crab," so called from its sluggish appearance. It is covered with a kind of pile or hair. The fourth and fifth pairs of legs are very short, and close-pressed against the carapace.

Several species of Mediterranean and American crustacea may also be seen alive in one or other of these tanks of our public aquaria.

It is only within a few years that the very large and widely distributed family of marine animals known as "barnacles" (*Cirripedia*) have been proved to belong to the same order as crabs and lobsters, and therefore to be veritable crustaceans. Darwin's monograph of these interesting creatures has placed them in a new light. Unlike the evolution of many animals, which begin in a simple way, and gradually pass through embryonic stages to a more complex (as the lobster, for example), the barnacles are actually more highly organised when they are young, free-swimming, and crustacean-like, than when they have reached the adult condition. Their life-history is a retrogradation, zoologically speaking, in order the better to adapt them to the very peculiar habits of life which we find them affecting. We can group the *Cirripedia* into two natural divisions, stalked and sessile, of which the stalked barnacle (*Lepas anatifera*), and the sessile "acorn" barnacle, we find so uncomfortably covering seaside rocks, are relative examples. The beautiful plumes or gills protruded from

Fig. 206.

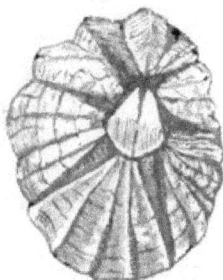

Sessile Acorn Barnacle (*Balanus porcatus*).

the semi-opened calcareous valves of the former, are well known. These constantly sweep the water in search of fresh oxygen and food. When the stalked species (so called from the flexible muscular tube, often of great length, to the free end of which the body of the barnacle is attached) have ceased leading a free-swimming life, they in reality settle down

Fig. 207.

Fig. 208.

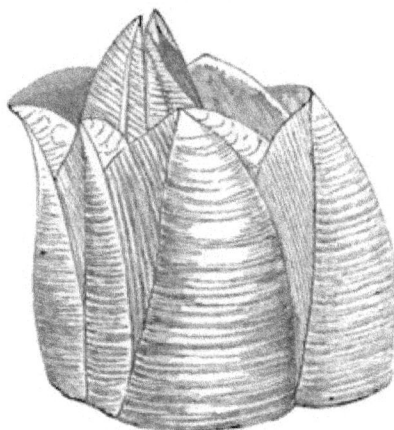

Balanus hameri.

Scalpellum vulgare.

by their heads. The feet then produce the feathery gills, and are hereafter employed in sweeping the water to and fro, instead of swimming or walking. In reality there is not much modification here, for if the legs of a lobster be plucked off sharply, there

will be found adhering to the point of attachment
a fringed oval object, which is part of a gill. Even
in the lobsters, therefore, foot-locomotion is also
partly subservient to breathing purposes. We have
several species of both stalked and unstalked bar-

Fig. 209.

Goose Tree (*Anseres arborei*).
From the ' Cosmographia Universalis ' of Munster.

nacles, of which *Scalpellum vulgare* (usually found
attached to the base of those corallines called "lobster's horn") are abundant among the former, and
Balanus hameri not unfrequent among the latter.

The young of *Scalpellum* pass through very similar

changes to those of the larvæ of *Lepas*. Our readers
will remember the *Lepas* as that which is usually
attached in thick clusters to old wreck and drift
wood These are difficult to keep in aquaria,
perhaps on account of their open-sea habits.   Mr.
Lloyd, however, managed to keep some alive at the
Crystal Palace for nearly six months on a floating

Fig. 210.

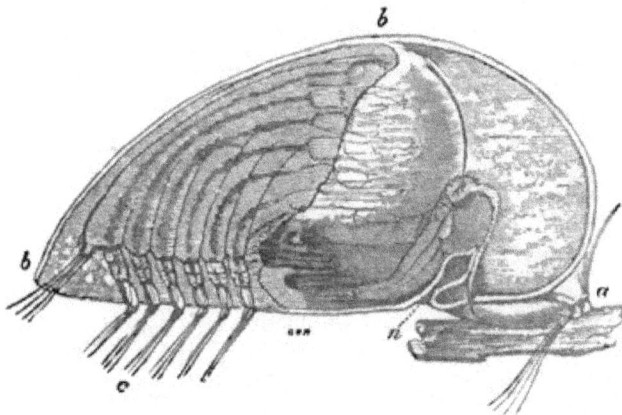

Larva of *Lepas Australis* in its last stage of development.   *a*. Antennæ,
with sucking disks. *b*. Carapace. *c*. Natatory legs.

bottle, found at Bridport.   This species usually goes
by the common name of the "goose barnacle," from
a very old notion (prevalent even among naturalists
two hundred and fifty years ago), that from the shells
the goose called the "barnacle" was produced.   It
was further believed to be borne by a peculiar kind
of marine shrub, which was called the "goose tree"

(*Anseres arborei*). Old Gerarde speaks most deci-
dedly of having witnessed the whole process of development ! The story is much older than his period,
and was told and illustrated by
Sebastian Munster, in his ' Cosmographia Universalis,' as long ago
as 1572.

Fig. 211.

The sessile barnacles appear to
be more easily acclimatised in
aquaria than the stalked. The
latter are usually drifted about at
sea, and possibly miss this artificial
mode of aeration when confined in
a tank.

The star-fishes, sea-urchins, and
marine worms, like the *crustacea*,
are capital adjuncts to what we may
call the " still life " of an aquarium.
They fill up as detail, make a good
fore or background, and intensify
the interest always felt in seeing ob-

Young Cirripede
(enlarged) immedi-
ately after moulting
the pupal carapace
and assuming its natu-
ral position. *a*. An-
tennæ. *c*. Natatory
unmoulted legs.

jects of which we have heard or read alive for the first
time. The star-fishes and sea-urchins are nearly related, in spite of their apparent unlikeness, both being
grouped into an order called *Echinodermata*, or " spiny
skinned." Both move by means of suckers, which
are worked by a wonderful hydraulic apparatus from
within the test or shell. In the sea-urchin the shell is
made up of at least six hundred pieces, mosaicked

together. Of these, five rows are perforated, and
through these perforations minute suckers are protruded. They can be elongated when injected with

Fig. 212.

Echinus climbing up side of an aquarium by means of its ambulatory
suckers.

water, which is strained off from the sea and admitted to the interior of the sea-urchin through a
special plate called the "madreporiform tubercle." All
these rows of feet protrude through the needle-like

Fig. 213.

A. Upper surface of Star-fish. B. Under surface of ditto, showing
sucking feet.

double rows of holes, which may be seen in any one
of the dead tests of the several species of sea-urchin

which frequent our own shores. The spines which have earned for this group the popular name of "Sea-urchin," are attached to minute rounded tubercles, on the plan of a "ball-and-socket joint," and can there-fore move about in all directions. But the *ambulacral* or sucking-feet can be protruded even beyond them, and thus the *Echinus*, as well as the star-fish, can

Fig. 214.

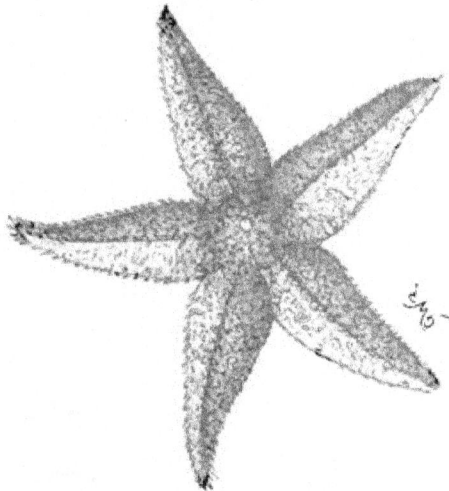

" Five-fingered " Star-fish (*Uraster rubens*).

glide horizontally, vertically, or even on surfaces over-head, with a quiet, ghost-like motion ; all the suckers being used to *warp* the body along.   In the star-fishes (with the exception of the brittle stars, which have no "water-vascular system," as the mechanism of suck-ing feet is called, but move about by entwining their lime-plated, snake-like arms) the sucking feet are

placed in rows underneath, whilst the sea water is admitted by a filtering plate situated on the upper surface. The Crystal Palace Aquarium has a large
number of species of this interesting order in the

Fig. 215.

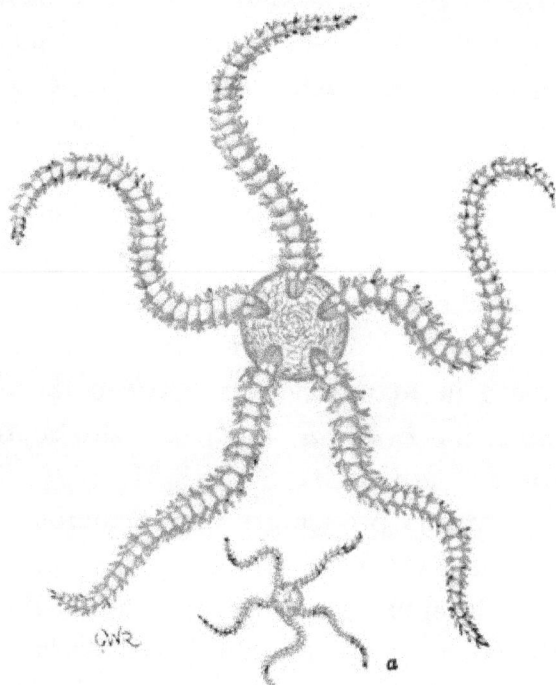

Ophiocoma neglecta. *a.* Natural size.

living state, among which we may mention *Uraster*,
or "five-fingered star-fish"; *Solaster*, or "twelve-rayed
star-fish;" cushion stars (*Goniaster*), "bird's foot"
star-fish (*Palmipes*), *Cribella*, *Comatula* (a species
allied to the crinoids, and which is indeed a crinoid
in its larval state); *Asterina*, &c. Among the brittle-

stars (so called from the readiness with which they
detach their arms when captured), we may there be-
hold *Luidia fragilisissima* (which has never been
kept healthily alive before), *Ophiocoma*, *Ophiura*, &c.
The "five-fingered" star-fish (*Uraster rubens*) is very
destructive to oysters, and may be seen in aquaria
devouring these delicious mollusca by insinuating
its bladder-like stomach between the two shells.
These, in some mysterious way, soon succumb to the
star-fish, and open as if the oyster were completely
paralysed ; so that the soft body is not long before it
becomes the prey of the persevering star-fish.

The "sea-cucumbers" (*Holothuriadæ*), as well as the
Sipunculidæ—a group of marine transitionary worms,
intermediate in many respects between the annelids
and some of the *Echinodermata*—are also represented
in most of our aquaria.

The sea worms proper are very numerous. Some
of them are too rapacious to be kept, except by themselves, and do injury to the rest of the inhabitants of
the tank. Among these is the black Borlase's worm
(*Nemertes Borlasii*), several yards in length, which
stretches from one end of the tank to the other, like so
many foldings of ribbon. The *Terebella* is an interesting object, half-way between naked worms and those
which live in limey tubes of their own secreting. The
Terebella simply forms an outer protection by cement-
ing grains of sand together, after the manner of a
caddis-worm. *Serpulæ* (which may be seen in abun-

dance on the backs of old oyster shells, &c.) secrete
a calcareous tube, as compact as the valve of a
mollusc. The crimson plumes are protruded from
the upper part, and these sweep the water for fresh

Fig. 216.   Fig. 217.

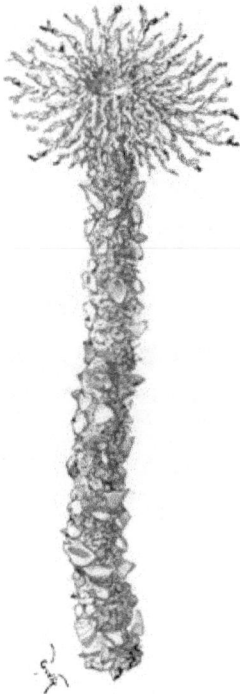

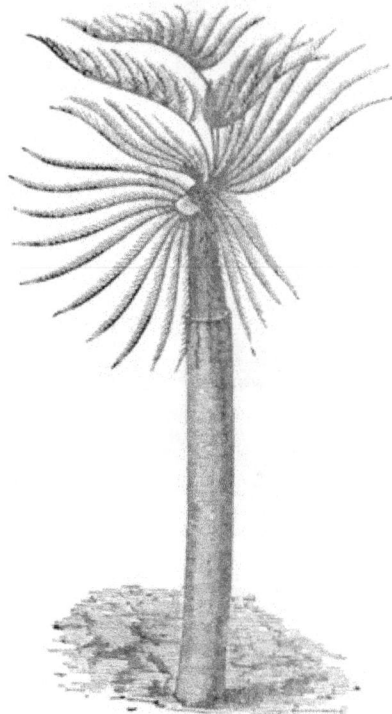

Tube of Terebella.   *Sabella unispira.*

air. In colour they are very contrasted with the
larger and still more feather-like gills of the *Sabella*
—a large marine worm, to be seen in marine tanks
burrowing in sand, or affixed to stones—which are
generally of a dull grey, although they are sometimes

of a bright colour.   This marine worm also constructs
its long and elegantly convoluted tube by means of
cemented grains of sand.

In addition to these our public aquaria have *Sabel-*

Fig. 218.

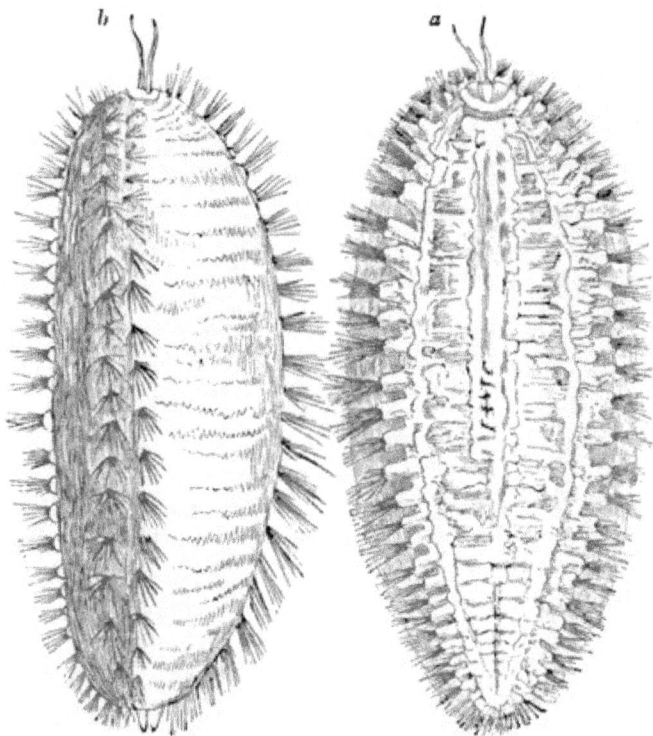

"Sea-mouse" (*Aphrodita aculeata*).   *a*. Under surface.   *b*. Side view.

*laria, Eunice, Aphrodita, Polynoe, Nereis, Pontobdella,
Arenicola* (or "lob-worm"), *Protula, Spirorbis, Phlly-
doce* (a worm of attractive appearance), *Eunice,
Pectinaria, Othonia,* &c.   These interesting creatures

have been made a speciality at the Crystal Palace, where the largest number of varieties may be seen alive. Of all these marine worms, however, certainly none is so attractive and so much sought after as the "sea-mouse" (*Aphrodita aculeata*). Its metallic lustre of green, blue, and yellow hairs, shining like those of the peacock's tail, would make it attractive by whatsoever name it might be called. Few people can

Fig. 219.

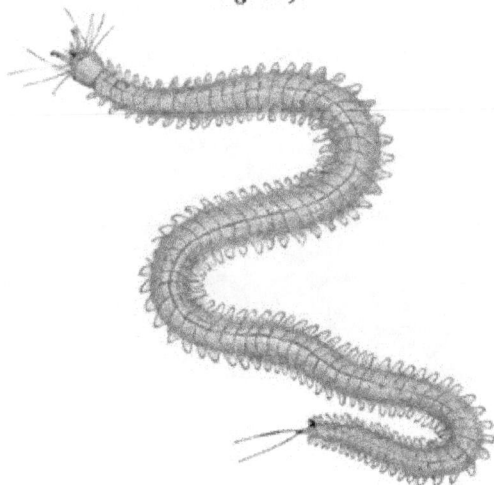

Nereis.

believe that this beautiful and not uncommon creature is a marine *worm*. Its back is covered with plates, underneath which are the breathing organs or *gills*. The plates are covered with the iridescent bristles, which, although beautiful even when the *Aphrodita* is cast up as a dead object at high water, are exceedingly lovely when it is seen alive and healthy in an aquarium. There can be no wonder, therefore, that it is sought after, and consequently exhibited in our public aquaria. Others of the non-tubed worms, the *Nereis*, for example, although not so brilliantly coloured, are very graceful and pretty marine creatures. The habits of some of the prettiest, however, belie their lovely appearance : for they are not unfrequently those of the well-clad stage ruffian who struts about in garments which have been obtained by means of murder and robbery. Not a few of the "errant," or wandering worms, live by stealthily preying upon objects actually more highly organised than themselves.

CHAPTER XVI.

SEA-ANEMONES AND OTHER ZOOPHYTES, ETC., OF MARINE AQUARIA.

EVEN before the days of large public aquaria, Gosse, Lankester, and others had taught us the ease with which those charming, flower-like objects called "sea-anemones" could be kept alive in vessels of sea water. No flowers in full bloom exceed them in colour or graceful shape, whilst in them there is superadded the extra interest which *life* gives to any object. These sea-anemones (*Actinia*) have been studied and observed more than any other group of marine animals. They are easy to keep alive, with certain necessary precautions, but require some little feeding. So flower-like are they that even insects are occasionally deceived by their floral appearance. Sir John Lubbock and others have recently shown how wonderfully co-related flowers and insects are ; and if it had been a less trustworthy observer than the late Jonathan Couch who related the following, it would have been difficult to believe that an insect so intelligent as a bee could make so gross a mistake as to take a sea-anemone for an open flower! Mr. Couch states that he saw an expanded "crass"

U

or "dahlia" wartlet anemone (*Tealia crassicornis*),
which was just covered by a film of sea water in a
rock-pool. It looked very attractive, and whilst he
was admiring its beauty a bee buzzed straight into
the embraces of the "crass," mistaking the tentacles
for petals, and paying for the error with its life, for
the remorseless fingers clutched it in their grasp and
transferred it to the ready stomach !

Fig. 220.

The Dahlia Wartlet (*Tealia crassicornis*).

The bright colours and elegant shapes of sea-anemones have caused them to be much sought after,
both by amateur aquarium-keepers and the managers
of our public institutions. About thirty species of
them and their allies may be seen living in the tanks
of the Crystal Palace, where they are usually fed on

a diet of chopped mussels, conveyed to them by
wooden forceps. In the sea they are thankful for

Fig. 221.

Plumose Anemone (*Actinoloba dianthus*).

any organic waif or stray that may come within their
reach. In the deeper tanks, where certain sea-anemones are placed, when live shrimps are turned into

U 2

the water the anemones occasionally catch them. Some species can slowly move from one place to another by means of the base of the column, after the manner of a snail crawling.   Perhaps the most noble looking of all this group of animals is the plumose anemone (*Actinoloba dianthus*), whose crown of feathered tentacless entitle it to the name of "sea-pink."   Its base is more expanded than is usual with other sea-anemones.   The species called the

Fig. 222.

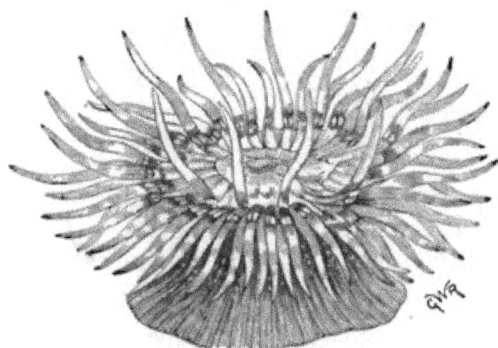

The Orange-disked Anemone (*Sagartia venusta*).

"orange-disk anemone" (*Sagartia venusta*) is named from its prevailing colour ; the tentacles, however, being a pure white.   The genus *Sagartia* was so named by Gosse on account of their possessing peculiar kinds of darts, stored-up in "nettling cells," which they can protrude so as to benumb and overcome their prey.   They are the most active of all the sea-anemones, and bear the most variegated colours, so that some species are indicated by names of colour. Thus we have the snowy anemone (*Sagartia nivea*), the rosy anemone (*Sagartia rosea*), &c. Their habits of life, also, seem more various than those of any other genus. One is called the "cave-anemone" (*Sagartia troglodytes*), from its habit of burrowing. Another well-

Fig. 223.

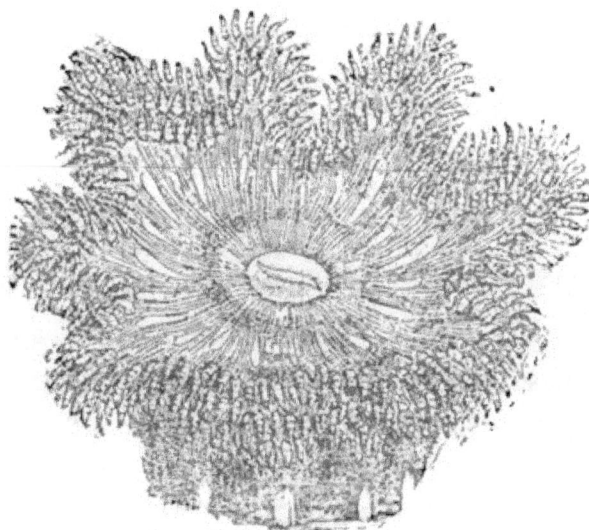

The Daisy Anemone (*Sagartia bellis*).

known species of *Sagartia* is called the "daisy"; whilst a very interesting species has formed some sort of an alliance with hermit crabs, so that it is usually found covering the shell in which one of these crustaceans has taken up its abode, and is therefore called the parasitic anemone (*Sagartia parasitica*). This is not the only sea-anemone which has contracted such a singular friendship. We have a pink-spotted one called the "cloaklet" (*Adamsia palliata*), so-called because it has no "column," and therefore spreads itself over the shell like a mantle. The attachment of these anemones for the hermit crabs is mutual, insomuch that when the latter change their shells for a larger, just before moulting, they will remove their zoophytal companions to their new abodes. Van Beneden has termed this strange association "commensalism," or "messmateship."

Fig. 224.

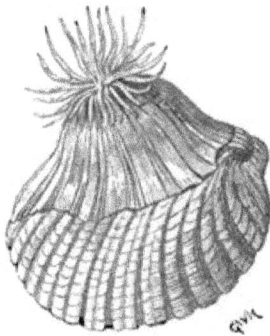

Parasitic Anemone (*Sagartia parasitica*).

Some of our public aquaria exhibit these species of sea-anemones in the position we have just described.

The commonest anemone is that known as the "beadlet" (*Actinia mesembryanthemum*), whose specific name is vulgarly shortened to "mes," as that of the "dahlia wartlet" is to "crass." It is a pretty creature, and cannot be too common; for its red tentacles, with their ring of turquoise-like spots, are exquisitely flower-like. It is also very hardy and long-lived, one specimen in Scotland having attained the authenticated age of more than *forty* years. All the sea-anemones, however, seem to be very sensitive at the base or foot of the column. If this be hurt they

die or are sickly, so that the best way is either to
peel them gently off the rock they are attached to,
or else to chip off the rock fragment to which a
specimen is adhering, so as to bring both away.

Fig. 225.   Fig. 226.

The Beadlet.
(*Actinia mesembryanthemum*).

The Opelet (*Anthea cereus*).

Corals differ from sea-anemones in having an interior hard skeleton, formed of lime. This is part
and parcel of the animal, just as our bones are of our
own bodies. If our readers can imagine the radiated
walls seen in a cross-section of the interior of the
body of a common sea-anemone to have the power of
secreting lime, they will understand how the radiating
septa of corals are formed. When a coral zoophyte
is dead, the anemone-like flesh and tentacles which
covered it decompose, and thus there is left behind
the pretty, hard, whitish body we call "coral." The
red coral of commerce, so valued in jewellery, is a
tinted calcareous skeleton secreted by another group
of animals (not the true coral animal), which belongs
to the same class as the "sea-fans" (*Gorgonidæ*). In

the compound corals, which are chiefly reef-building, the individual zoophytes are usually smaller, as in the *Millepores.* Two of our British genera are *Caryo-*

Fig. 227.

Balanophyllia regia.

Fig. 228.

Caryophyllia.

phyllia and *Balanophyllia.* They may be seen alive in nearly all our aquaria, and should be examined in order to gain a clear idea of what a coral animal really is like. It is very difficult to convey the correct idea—people have called them "insects," and ima-

gined they built up the
hard limey *coral* as bees
do their combs; whereas
we have seen it is really
part of themselves, covered
over with flesh (except the
base in very old specimens)
when the animals are alive.
Balanophyllia verrucosa is a
species most generally seen.
It is of a bright orange
colour, and abounds in the
Mediterranean. Several
living Italian species are
exhibited at the Crystal
Palace Aquarium.

The dead man's fingers
(*Alcyonium digitatum*) is an
object well known to fishermen, both by this name and
that of "cow's paps," &c.
It is dredged up from
deepish water, although
frequently found stranded
between tides. It is usually
attached by its base, the
body swelling upwards, and
covered with papillæ when
taken out of the water.
The general body colour

Fig. 229.

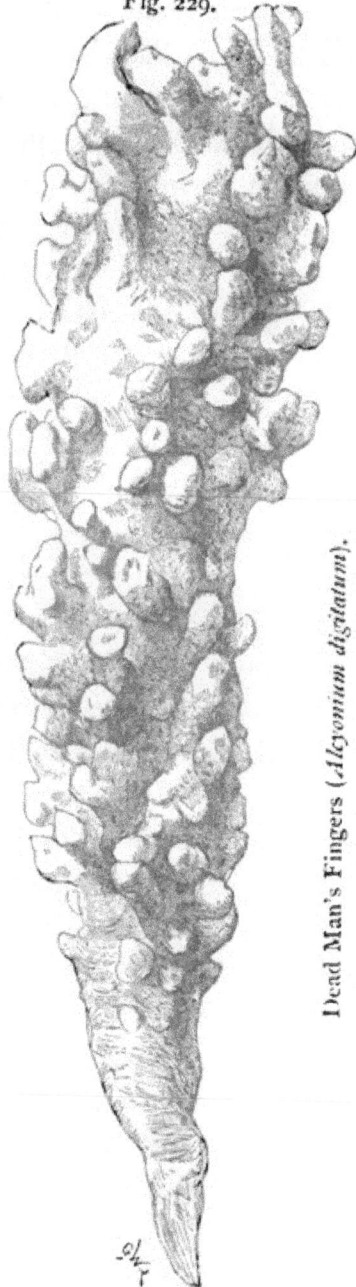

Dead Man's Fingers (*Alcyonium digitatum*).

varies from yellowish-white to orange-red. Each of
the little papillæ, when the *Alcyonium* is alive, expands into a lovely flower-like animal, so that the
fleshy-looking mass called dead man's fingers is in
reality a colony of zoophytes. The flesh is braced by
the distribution through it of a number of spicules of
lime. At the Crystal Palace, Brighton, and elsewhere, these objects are kept alive, and the action of
the fringed petal-like tentacles of the zoophytes may
be witnessed through a magnifying glass. *Alcyonium*
is one of the sea-fans (*Gorgonidæ*), the dried, horny
skeletons of which may often be seen in mariners'
houses, or in museums. They usually possess a dark
horny axis covered with a red, orange, or pinkish
skin, when dried. These objects are also colonies of
zoophytes, of whose hard horny skeleton we have
been speaking. When alive this skeleton is covered
with flesh, and out of the latter there spring buds,
just as the bark covers a tree, and allows buds to
burst through it. We give the figure of a well-known
form called *Isis hippuris*, in which the relation of the
skeleton to the external flesh and zoophytes is at
once seen. *Isis* is found on the east coast of Scotland, and the Orkney Islands.

We have several species of British gorgonias, of
which perhaps *Gorgonia verrucosa*, and *G. flabellum*
are the largest and handsomest. These have been
kept alive at the Crystal Palace Aquarium for a short
time, but there is a difficulty in knowing how to feed

them, so that they soon die. In the "fan" gorgonia (*G. flabellum*) the branches run into each other, so as to

Fig. 230.

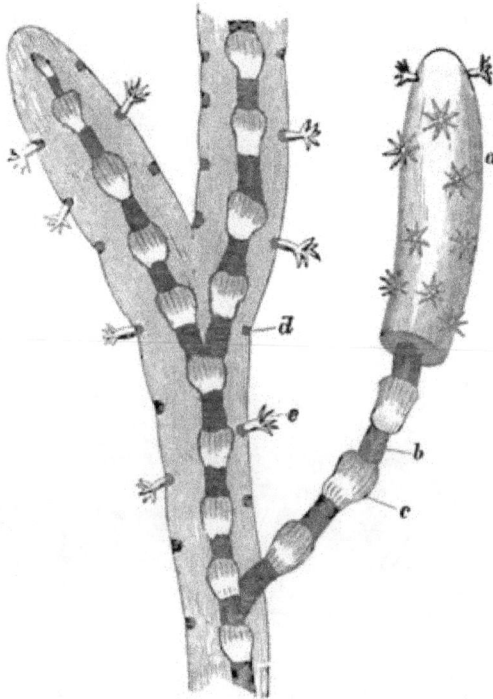

Isis hippuris × 5 (vertical section), showing flesh, with zoophytes investing internal horny axis. *a*, external appearance.

unite. In Fig. 232, we give a slightly magnified illustration of part of one of the branches. The circular pits, one in the common flesh (*Cœnosarc*) which invests the skeleton, and one occupied by the individual polypes. All the British species of this interesting order are exceedingly pretty; although not to be compared with the gorgeous appearance of the sea-

fans of tropical seas—so graphically described by
Schleiden. We should be glad to see more of them

Fig. 231.

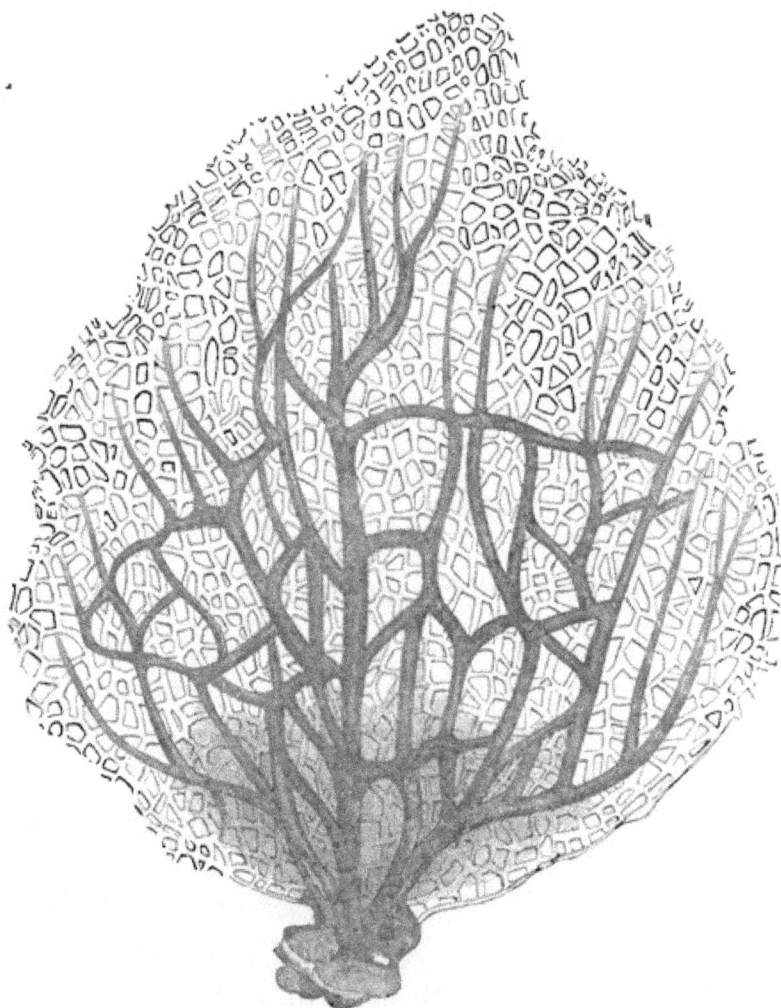

The Common Sea-fan (*Gorgonia flabellum*).

kept in our public aquaria, as the dried specimens
are very common and well-known ; although most

people, even when generally well educated, know
little of what they are. If shown in a tank in the

Fig. 232.

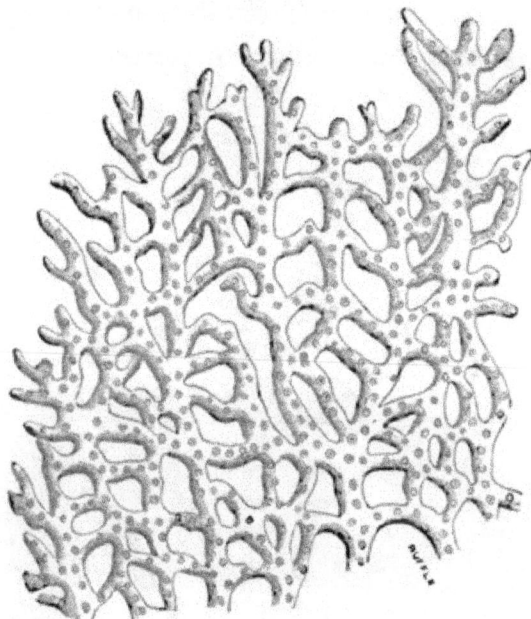

Small portion of *Gorgonia flabellum*, showing pits in cœnosarc,
occupied by Polypes.

living state, this ignorance would vanish at their
first sight. In *Gorgonia verticillaria* the polypes are
arranged around the axis like the leaves of such a
verticellate plant as the common " goose grass " or
" cleavers," The species called *pinnata*, is not very
fan-like in shape, although there needs little effort
to see that it belongs to the sea-fans. We have
referred particularly to our British species because
we think it would be easier to domesticate them in
aquaria than foreign species. Some of them are not

rare off the southern coasts of England, and may easily be obtained alive.

The smaller zoophytes are too inconspicuous to be kept in tanks for public exhibition, although many naturalists keep them alive for purposes of research.

Fig. 233.

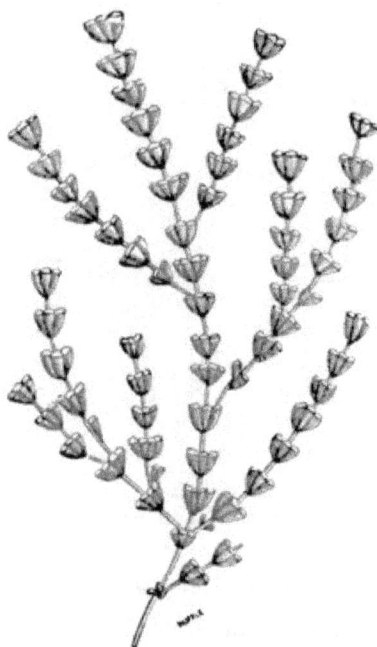

Gorgonia verticillaria.

No other group of marine animals is so interesting, especially since it has been discovered that the minute *Campanularians* give rise to large jelly-fish-like progeny.   Indeed, many of the so-called me-dusoid jelly-fishes are but the larval conditions of

Fig. 234.

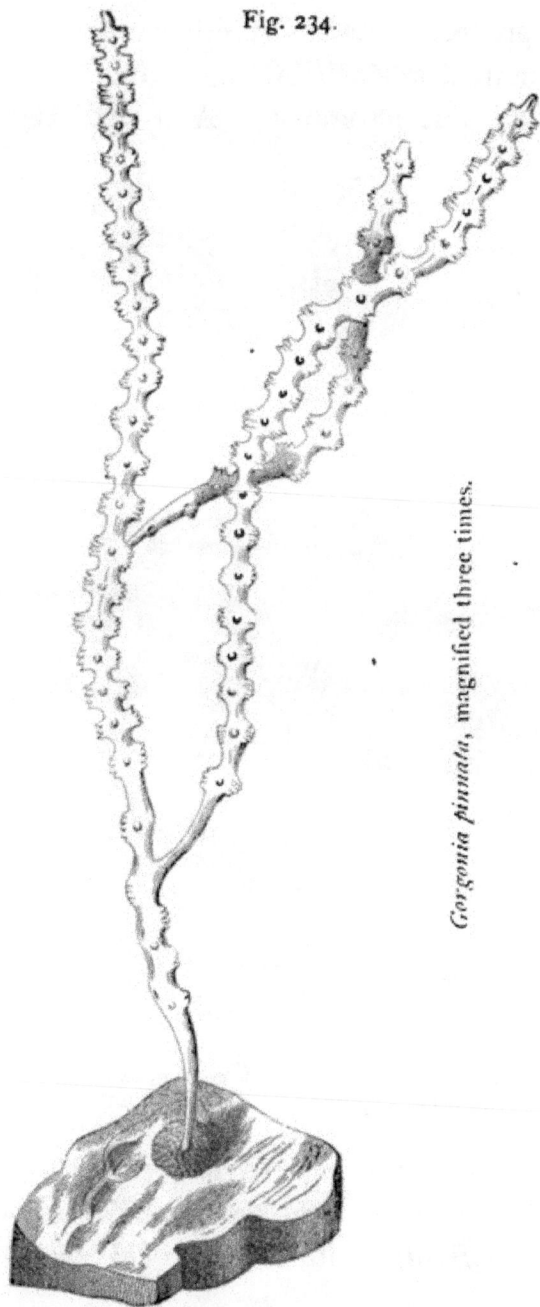

Gorgonia pinnata, magnified three times.

hydroid zoophytes. Few of these campanularians
are more than half an inch in length, and are rarely
noticed by any but naturalists. And yet they are

Fig. 235.

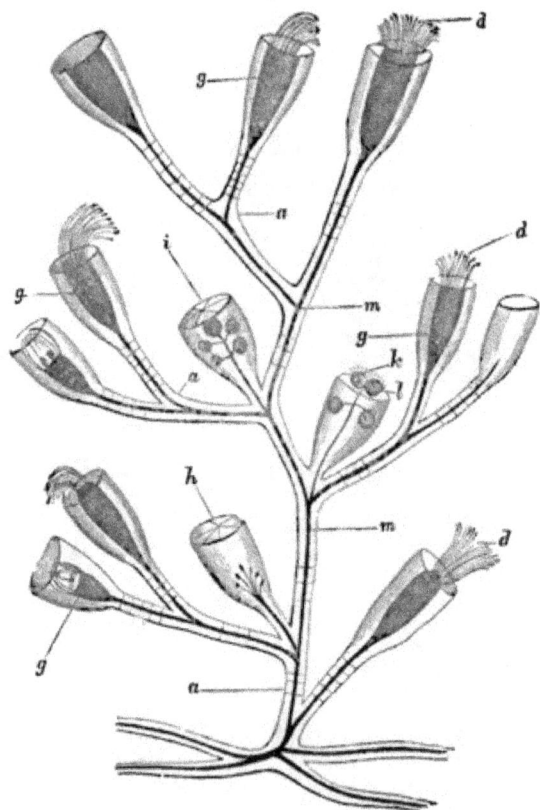

Diagrammatic sketch of *Campanularia dichotoma*, showing common
flesh (*m*), cilia of each zoophyte (*d*), body of zoophyte (*g*), &c.

covered with the minute cups of living zoophytes
(Fig. 235, *d, g, h,* &c.). The "oaten-pipe" coralline
(*Tubularia indivisa*), with its coronal of scarlet

tentacles, waving from out the horny "pipes," is a well-known object, and this may be seen at the Crystal Palace, together with the "lobster's horn"

Fig. 236.

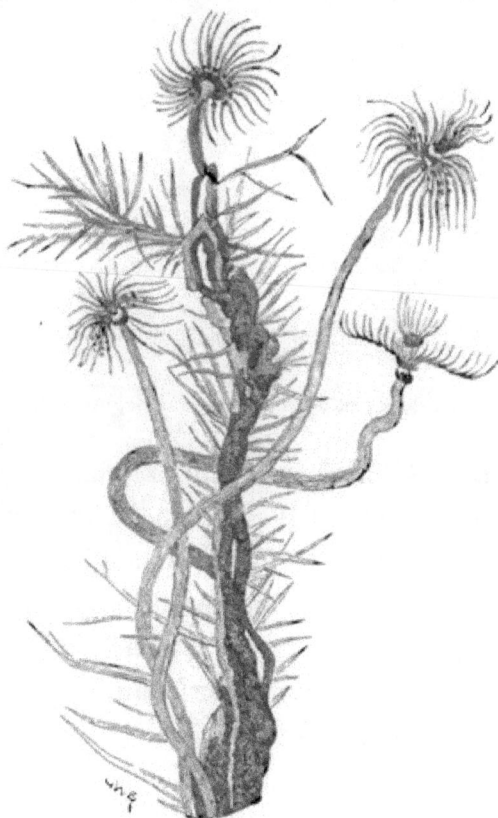

Oaten-pipe Coralline (*Tubularia indivisa*).

coralline (*Antennularia antennina*), and the true "sea-fir (*Sertularia abietina*, Fig. 183), all of them hydroid zoophytes. It is difficult to preserve jelly fishes,

X

although *Cydippe* has been kept alive a little time, and we have seen the larval *Medusæ* at Brighton and the Crystal Palace, not bigger than a pin's head, which had been hatched in the tanks. The pretty little *Hydra-tuba* may also be searched for and found

Fig. 237.

Hydra-tuba, in various stages of segmentation.

in these places; and in them, again, we have evidence of how "jelly fishes" can be formed from an object less than half an inch in height, by the "segmentation," or splitting up into free parts, of saucer-like objects, which turn over into umbrella-like disks, and then swim away, as utterly unlike their parents as any two animals can well be.

Sponges will always be difficult objects to keep alive under artificial conditions, although we have seen that the most delicate of all of them, the fresh-water sponge (*Spongilla fluviatilis*) has succumbed to the will and care of man in this respect. Several of our commonest British sponges are kept alive at Brighton, Manchester, and especially at Sydenham, where they grow so well that they form masses as big as a man's head. But some of them are not attractive-looking creatures ; and, indeed, few people would imagine them to be animals at all. The *spongy* object which we know as sponge, is only the inner skeleton ; just as *coral* is of another class of marine animals when alive. This substance is covered all over with a transparent gelatinous flesh, called *sarcode*, which lines every pore and every aperture. In and out of these currents of sea water perpetually flow, induced by the action of the eyelash-like processes (*cilia*) with which the surface of the flesh lining such hollows is covered. Among them we have the "crumb-of-bread" sponge (*Halichondria panicea,* Fig. 238), the *Chalina oculata,* Fig. 239 (largest of our British sponges), the *Leucosalenia, Tethea, Grantia, Hymeniacidon, Cliona* — the latter is really a boring sponge, and burrows the holes we may see covering the surfaces of old oyster-shells. Some of these sponges are of a brilliant yellow or crimson colour, and mat the surfaces of rockwork, &c., having been introduced into the sea water as "gemmules," and developed by circumstances into their present attractive and useful conditions.

No doubt the marine aquarium might be converted into a nursery for microscopic objects as well as the fresh water. There is no end to the variety of the lower forms of life inhabiting the sea. They are the oldest of all the animal kingdom, having been in

existence since perhaps before the distant Laurentian
epoch of which geology speaks.   They have experienced all the changes which we know the ocean
beds must have experienced—have repeatedly witnessed the conversion of continent into sea, and

Fig. 238.

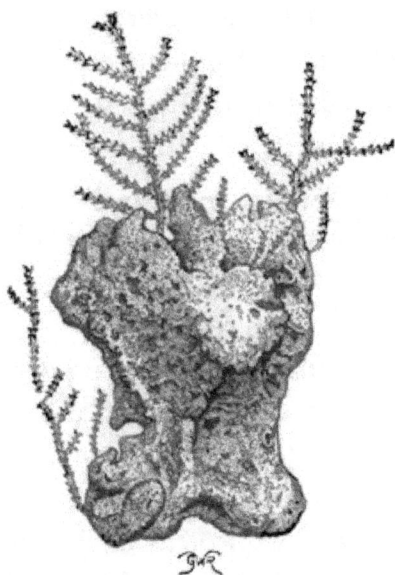

" Crumb-of-bread " Sponge (*Halichondria panicea*).

ocean floor into dry land and even mountain chain.
The mineral remains of some of the ancestors of
living forms have taken an active part in the building
up of rock masses of every geological period.   We
cannot wonder, therefore, that the number of species
of these early and unpretending forms of life should

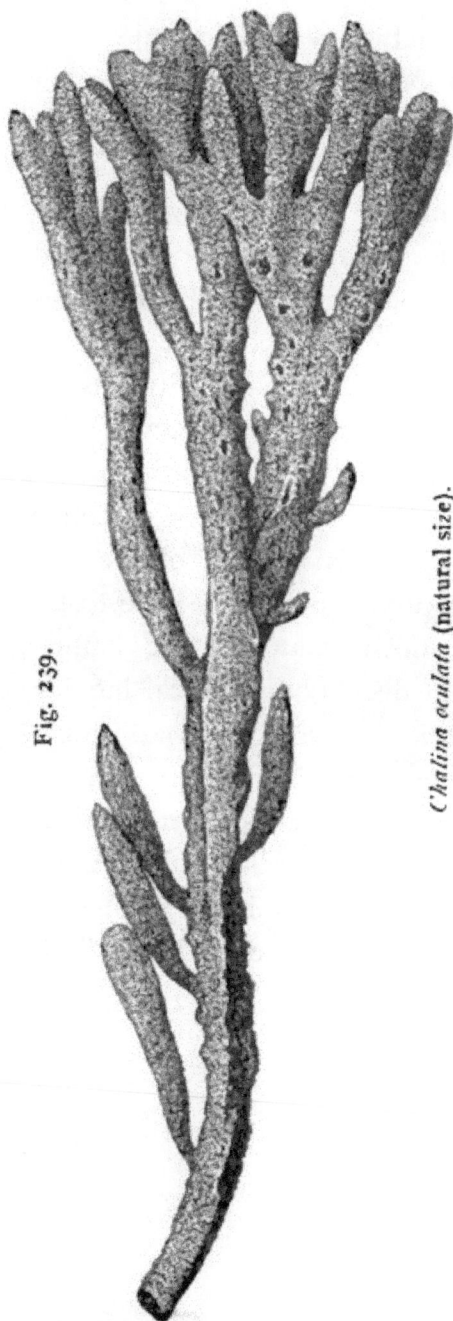

Fig. 239.

Chalina oculata (natural size).

be so great, or that they should enjoy so cosmopolitan a distribution. There is no reason why more
naturalists should not cultivate the humblest as well
as the most highly-developed types of marine life,
and keep living marine diatoms, foraminiferæ, nocti-
lucæ, and sponges. Our rock-pools support them,
and it is a sad proof of how little we yet know of
the natural conditions of such little spots as these,
if we cannot keep them artificially alive also.

That aquaria are still in the infancy of their development we do not doubt, any more than that they
will administer to the growing love of animated nature which is the especial feature of the intellectual
culture of our century. Here will have to be fought
out and hunted down many of the embryological
questions to which deep philosophical inquiry is now
attaching such great importance. And, whilst aquaria
may in this manner be useful to true science, they
will not be less so to unscientific people, in revealing
to them at a glance the shapes, habits, and natures
of creatures they had never heard of before, so as
thus to form a practical education all the more
valuable because those who learn are for the time
unaware of its importance.

INDEX.

LONDON: PRINTED BY WILLIAM CLOWES AND SONS, STAMFORD STREET
AND CHARING CROSS.